INTRODUCTORY CHEMISTRY

Edited by
Jean Annerie Hernandez

Delve Publishing

Introductory Chemistry

Edited by Jean Annerie Hernandez

Edition 2018

Delve Publishing

Published by
Delve Publishing LLC
708 3rd Avenue, 6th Floor
New York
NY 10801
United States of America
www.delvepublishing.com

ISBN: 978-1-68095-724-2 (Hardback)

Printed in United States of America

Library of Congress Control Number: 2016920404

About the Editor

Jean Annerie Hernandez

Jean Annerie took up a Bachelor's degree in Chemistry in Central Luzon State University before she decided to fulfill her lifelong dream of becoming a creative writer and editor. She has ghostwritten and copyedited fictional and non-fictional books for three years, and is currently working on various client-based projects in Singapore, China and USA

Table of Contents

List of Abbreviations xi
Preface xiii

1 INTRODUCTION CHEMISTRY

Introduction 1
Etymology 2
Definition 2
History 3
Chemistry as Science 4
Chemical Structure 6
Principles of Modern Chemistry 7
Matter 8
Element 9
Compound 10
Molecule 10
Substance and Mixture 12
Mole and Amount of Substance 14
Phase 14
Bonding 15
Energy 16
Reaction 18
Ions and Salts 19
Acidity and Basicity 20
Redox 21
Equilibrium 21
Chemical Laws 21
Practice 22
Subdisciplines 22
Chemical Industry 24
References 24

2 ATOMIC THEORY

Introduction 30
History 31
Atomism 38
Reductionism 38
Greek Atomism 39
Indian Atomism 41
Islamic Atomism 44
Medieval European Speculations 45
Atomic Renaissance 46
Atomic Theory 48
Atomic Theory Controversy 49
Kinetic Theory of Gases 49
Assumptions 50
Properties 51
Temperature and Kinetic Energy 52
Collisions with Container 54
Speed of Molecules 54
Physicalism 54
Definition of Physical 55
Reductionism and Emergentism 58
A Priori Versus a Posteriori Physicalism 59
Other Views 60
References 61

3 MATTER

Introduction 67
Comparison with Mass 69
Definition 70
Based on Mass, Volume, and Space 70
Based on Atoms 70
Based on Protons, Neutrons and Electrons 70
Based on Quarks and Leptons 70
Based on Theories of Relativity 72
Structure 73
Quarks 73
Phases 76
Antimatter 77
Other Types 77
Dark Matter 78

Dark Energy 78
Exotic Matter 79
Historical Development 79
Antiquity (C. 610 Bc–C. 322 Bc) 79
Summary 83
References 83

4 CHEMICAL COMPOUND

Introduction 91
Definitions 92
Elementary Concepts 93
Comparison to Mixtures 93
Formula 94
Phases and Thermal Properties 94
Chemical Element 94
Description 97
Atomic Number 98
Isotopes 99
Isotopic Mass and Atomic Mass 99
Chemically Pure and Isotopically Pure 100
Allotropes 100
Properties 101
General Properties 101
States of Matter 101
Melting and Boiling Points 101
Densities 102
Crystal Structures 102
Occurrence and Origin on Earth 102
The Periodic Table 103
Nomenclature And Symbols 104
Origin of the Elements 106
Abundance 108
Discovery and Recognition of Various Elements 113
Recently Discovered Elements 114
Chemical Revolution 114
Primary Factors 115
Precision Instruments 116
Antoine Lavoisier 116
Chemical Structure 117
Chemical Nomenclature 118
Aims of Chemical Nomenclature 118

Differing Aims of Chemical Nomenclature and Lexicography 119
Types of Nomenclature 119
Iupac Nomenclature of Inorganic Chemistry 120
System 120
References 121

5 CHEMICAL BOND

Introduction 127
Overview of Main Types of Chemical Bonds 128
History 130
Bonds in Chemical Formulas 131
Strong Chemical Bonds 131
Ionic Bonding 131
Covalent Bond 133
Single and Multiple Bonds 134
Coordinate Covalent Bond (Dipolar Bond) 135
Metallic Bonding 135
Intermolecular Bonding 135
Theories of Chemical Bonding 136
References 136

6 CHEMICAL REACTIONS

Introduction 139
History 141
Equations 142
Elementary Reactions 143
Chemical Equilibrium 144
Thermodynamics 144
Kinetics 145
Reaction Types 146
Four Basic Types 146
Oxidation and Reduction 148
Complexation 150
Acid-Base Reactions 150
Precipitation 151
Solid-State Reactions 152
Reactions at the Solid | Gas Interface 152
Photochemical Reactions 152
Catalysis 152
Reactions in Organic Chemistry 154

Biochemical Reactions 162
Applications 162
Monitoring 163
References 163

7 CHEMICAL EQUILIBRIUM

Introduction 167
Historical Introduction 167
Thermodynamics 170
Treatment of Activity 174
Effect of Temperature 177
Types of Equilibrium 177
Composition of A Mixture 178
Determination of Equilibrium Constants 181
Experimental Methods 181
Computational Methods 184
Equilibrium Constant 192
Types of Equilibrium Constants 193
Thermodynamics 198
Temperature Dependence 202
Pressure Dependence 204
Henderson–Hasselbalch Equation 205
For Bases 206
Derivation 206
Limitations 207
Conventions 208
Estimating Blood Ph 208
Michaelis–Menten Kinetics 209
Model 209
Applications 211
Derivation 211
Equilibrium Approximation 212
Quasi-Steady-State Approximation 213
Determination of Constants 214
Role of Substrate Unbinding 215
References 215

8 CHEMICAL COMPOSITION

Introduction 221
Composition of a Substance 221

Composition of a Mixture....222
Mole Fraction....222
Properties....223
Related Quantities....223
Spatial Variation and Gradient....224
Volume Fraction....224
Volume Percent....225
Mass Fraction (Chemistry)....225
Terminology....226
Properties....226
Related Quantities....227
Molality....228
Definition....228
Origin....228
Unit....228
Usage Considerations....229
Relation to Other Compositional Properties....229
Relation to Apparent (Molar) Properties....233
Molar Concentration....233
Definition....233
Units....234
Related Quantities....234
Properties....236
Formal Concentration....237
Equivalent Concentration....237
Unit Symbol N....237
Usage....238
Examples....238
Criticism....238
References....238

9 INTERMOLECULAR FORCE

Introduction....241
Dipole-Dipole Interactions....242
Van Der Waals Forces....243
Relative Strength of Forces....245
Effect on the Behavior of Gases....245
Quantum Mechanical Theories....246
References....246

Index....247

List of Abbreviations

Cl	Chlorine
IUPAC	International Union of Pure and Applied Chemistry
VSEPR	Valence Shell Electron Pair Repulsion Model
DNA	Deoxyribonucleic Acid Molecules
WMAP	Wilkinson Microwave Anisotropy Probe
CAS	Chemical Abstracts Service
STP	Standard Temperature and Pressure
LCAO	Linear Combination of Atomic Orbitals Molecular Orbital Method
ST	Specific Ion Theory
SI	International System of Units
HCl	Hydrogen Chloride

Preface

Everything you hear, see, smell, taste, and touch involves chemistry and chemicals (matter). And hearing, seeing, tasting, and touching all involve intricate series of chemical reactions and interactions in your body. With such an enormous range of topics, it is essential to know about chemistry at some level to understand the world around us. Commonly, chemistry is a science that deals with how materials form, their composition, various attributes, how they transform, and the energy they consume or release through these transformations. Chemistry relates to each and everything you can sense, from right down to the elemental level to massive complex structures. It deals with how things are constructed and formulated to how they break apart and the energy that is either absorbed or released in each of these processes. The study of the science of chemistry shows us, for example, how the table salt we eat every day is actually constructed of crystals formed out of sodium and chlorine, and why the Statue of Liberty, constructed out of metallic copper is now green. Chemistry is such a broad subject and one so full of detail that it is easy for a newcomer to find it somewhat overwhelming, if not intimidating. Chemists sometimes refer to matter as 'stuff', and indeed so it is. Matter is anything that has mass and occupies space. Which is to say, anything you can touch or hold. Common usage might have us believe that 'chemicals' are just those substances in laboratories or something that is not a natural substance. Far from it, chemists believe that everything is made of chemicals. Although there are countless types of matter all around us, this complexity is composed of various combinations of some 100 chemical elements. The names of some of these elements will be familiar to almost everyone. Elements such as hydrogen, chlorine, silver, and copper are part of our everyday knowledge. Far fewer people have heard of selenium or rubidium or hassium. Nevertheless, all matter is composed of various combinations of these basic elements. The wonder of chemistry is that when these basic particles are combined, they make something new and unique. And so it is with chemistry, understanding the basic properties of matter and learning how to predict and explain how they change when they react to form new substances is what chemistry and chemists are all about. This book is designed as a brief introduction to chemistry. The primary objective is to give you enough understanding of the fundamental concepts and language of chemistry.

Content Coverage

Chapter One presents an introduction to chemistry as the study atoms, which are the basic building blocks of matter, as well interactions between atoms. It also presents brief overview of subatomic particles, which are smaller than atoms, and these include things like protons, neutrons, and electrons.

Chapter Two presents a brief overview of atomic theory; a scientific theory of the nature of matter, which states that matter is composed of discrete units called atoms.

Chapter Three covers the theory of matter that exists in states (or phases): the classical solid, liquid, and gas; as well as the more exotic plasma, Bose–Einstein condensates, fermionic condensates, and quark–gluon plasma.

Chapter Four and *Five* elaborate on the study of chemical compounds and chemical bonds. Since opposite charges attract via a simple electromagnetic force, the negatively charged electrons that are orbiting the nucleus and the positively charged protons in the nucleus attract each other.

Chapters Six and *Seven* cover the process of chemical reactions and chemical equilibrium, respectively. In a chemical reaction, chemical equilibrium is the state in which both reactants and products are present in concentrations which have no further tendency to change with time. Usually, this state results when the forward reaction proceeds at the same rate as the reverse reaction.

Chapters Eight and *Nine* attempt to explain chemical composition and intermolecular forces between atoms, respectively.

CHAPTER 1

Introduction Chemistry

Chapter Contents

- **Introduction**
- **Etymology**
- **History**
- **Practice**

Chemistry is a physical science, and it is the study of the properties of and interactions between matter and energy. In other words, chemistry is a way to study the properties, characteristics, and physical and chemical changes of matter.

Matter is pretty important because it's anything that has mass and takes up space - basically, all of the 'stuff' that makes up our world! Chemists study atoms, which are the basic building blocks of matter, as well interactions between atoms.

They also study subatomic particles, which are smaller than atoms, and these include things like protons, neutrons, and electrons. Since everything on Earth is made of matter, and matter is made of atoms, you can see how this creates the overlap between chemistry and other sciences. You can't have 'stuff' to study if you don't have 'stuff' in the first place. In other words, matter really matters!

INTRODUCTION

Chemistry is a branch of physical science that studies the composition, structure, properties and change of matter... Chemistry includes topics such as the properties of individual atoms, how atoms form chemical bonds to create chemical compounds, the interactions of substances through intermolecular forces that give matter its general properties, and the interactions between substances through chemical reactions to form different substances.

Chemistry is sometimes called the central science because it bridges other natural sciences, including physics, geology and biology... For the differences between chemistry and physics see comparison of chemistry and physics..

Scholars disagree about the etymology of the word chemistry. The history of chemistry can be traced to alchemy, which had been practiced for several millennia in various parts of the world.

ETYMOLOGY

The word chemistry comes from alchemy, which referred to an earlier set of practices that encompassed elements of chemistry, metallurgy, philosophy, astrology, astronomy, mysticism and medicine. It is often seen as linked to the quest to turn lead or another common starting material into gold,. though in ancient times the study encompassed many of the questions of modern chemistry being defined as the study of the composition of waters, movement, growth, embodying, disembodying, drawing the spirits from bodies and bonding the spirits within bodies by the early 4th century Greek-Egyptian alchemist Zosimos.. An alchemist was called a 'chemist' in popular speech, and later the suffix "-ry" was added to this to describe the art of the chemist as "chemistry".

The modern word alchemy in turn is derived from the Arabic word al-kīmīā (الكيمياء). In origin, the term is borrowed from the Greek χημία or χημεία... This may have Egyptian origins since al-kīmīā is derived from the Greek χημία, which is in turn derived from the word Chemi or Kimi, which is the ancient name of Egypt in Egyptian.. Alternately, al-kīmīā may derive from χημεία, meaning "cast together"..

Definition

In retrospect, the definition of chemistry has changed over time, as new discoveries and theories add to the functionality of the science. The term "chymistry", in the view of noted scientist Robert Boyle in 1661, meant the subject of the material principles of mixed bodies.. In 1663 the chemist Christopher Glaser described "chymistry" as a scientific art, by which one learns to dissolve bodies, and draw from them the different substances on their composition, and how to unite them again, and exalt them to a higher perfection..

The 1730 definition of the word "chemistry", as used by Georg Ernst Stahl, meant the art of resolving mixed, compound, or aggregate bodies into their principles; and of composing such bodies from those principles.. In 1837, Jean-Baptiste Dumas considered the word "chemistry" to refer to the science concerned with the laws and effects of molecular forces.. This definition further evolved until, in 1947, it came to mean the science of substances: their structure, their properties, and the reactions that change them into other substances - a characterization accepted by Linus Pauling.. More recently, in 1998, Professor Raymond Chang broadened the definition of "chemistry" to mean the study of matter and the changes it undergoes..

HISTORY

Early civilizations, such as the Egyptians. Babylonians, Indians. amassed practical knowledge concerning the arts of metallurgy, pottery and dyes, but didn't develop a systematic theory.

A basic chemical hypothesis first emerged in Classical Greece with the theory of four elements as propounded definitively by Aristotle stating that fire, air, earth and water were the fundamental elements from which everything is formed as a combination. Greek atomism dates back to 440 BC, arising in works by philosophers such as Democritus and Epicurus. In 50 BC, the Roman philosopher Lucretius expanded upon the theory in his book De rerum natura (On The Nature of Things)... Unlike modern concepts of science, Greek atomism was purely philosophical in nature, with little concern for empirical observations and no concern for chemical experiments..

In the Hellenistic world the art of alchemy first proliferated, mingling magic and occultism into the study of natural substances with the ultimate goal of transmuting elements into gold and discovering the elixir of eternal life.. Work, particularly the development of distillation, continued in the early Byzantine period with the most famous practitioner being the 4th century Greek-Egyptian Zosimos of Panopolis.. Alchemy continued to be developed and practised throughout the Arab world after the Muslim conquests,. and from there, and from the Byzantine remnants,. diffused into medieval and Renaissance Europe through Latin translations. Some influential Muslim chemists, Abū al-Rayhān al-Bīrūnī,. Avicenna. and Al-Kindi refuted the theories of

Democritus' atomist philosophy was later adopted by Epicurus (341–270 BCE).

Chemistry as Science

Jābir ibn Hayyān (Geber), a Persian alchemist whose experimental research laid the foundations of chemistry.

development of the modern scientific method was slow and arduous, but an early scientific method for chemistry began emerging among early Muslim chemists, beginning with the 9th century Persian or Arabian chemist Jābir ibn Hayyān (known as "Geber" in Europe), who is sometimes referred to as "the father of chemistry"..... He introduced a systematic and experimental approach to scientific research based in the laboratory, in contrast to the ancient Greek and Egyptian alchemists whose works were largely allegorical and often unintelligble.. Under the influence of the new empirical methods propounded by Sir Francis Bacon and others, a group of chemists at Oxford, Robert Boyle, Robert Hooke and John Mayow began to reshape the old alchemical traditions into a scientific discipline. Boyle in particular is regarded as the founding father of chemistry due to his most important work, the classic chemistry text The Sceptical Chymist where the differentiation is made between the claims of alchemy and the empirical scientific discoveries of the new chemistry.. He formulated Boyle's law, rejected the classical "four elements" and proposed a mechanistic alternative of atoms and chemical reactions that could be subject to rigorous experiment..

The theory of phlogiston (a substance at the root of all combustion) was propounded by the German Georg Ernst Stahl in the early 18th century and was only overturned by the end of the century by the French chemist Antoine Lavoisier, the chemical analogue of

Newton in physics; who did more than any other to establish the new science on proper theoretical footing, by elucidating the principle of conservation of mass and developing a new system of chemical nomenclature used to this day..

Before his work, though, many important discoveries had been made, specifically relating to the nature of 'air' which was discovered to be composed of many different gases. The Scottish chemist Joseph Black (the first experimental chemist) and the Dutchman J. B. van Helmont discovered carbon dioxide, or what Black called 'fixed air' in 1754; Henry Cavendish discovered hydrogen and elucidated its properties and Joseph Priestley and, independently, Carl Wilhelm Scheele isolated pure oxygen.

Antoine-Laurent de Lavoisier is considered the "Father of Modern Chemistry"..

English scientist John Dalton proposed the modern theory of atoms; that all substances are composed of indivisible 'atoms' of matter and that different atoms have varying atomic weights.

The development of the electrochemical theory of chemical combinations occurred in the early 19th century as the result of the work of two scientists in particular, J. J. Berzelius and Humphry Davy, made possible by the prior invention of the voltaic pile by Alessandro Volta. Davy discovered nine new elements including the alkali metals by extracting them from their oxides with electric current..

British William Prout first proposed ordering all the elements by their atomic weight as all atoms had a weight that was an exact multiple of the atomic weight of hydrogen. J. A. R. Newlands devised an early table of elements, which was then developed into the modern periodic table of elements. in the 1860s by Dmitri Mendeleev and independently by several other scientists including Julius Lothar Meyer... The inert gases, later called the noble gases were discovered by William Ramsay in collaboration with Lord Rayleigh at the end of the century, thereby filling in the basic structure of the table.

In his periodic table, Dmitri Mendeleev predicted the existence of 7 new elements,. and placed all 60 elements known at the time in their correct places..

Organic chemistry was developed by Justus von Liebig and others, following Friedrich Wöhler's synthesis of urea which proved that living organisms were, in theory, reducible to chemistry.. Other crucial 19th century advances were; an understanding of valence bonding (Edward Frankland in 1852) and the application of thermodynamics to chemistry (J. W. Gibbs and Svante Arrhenius in the 1870s).

Chemical Structure

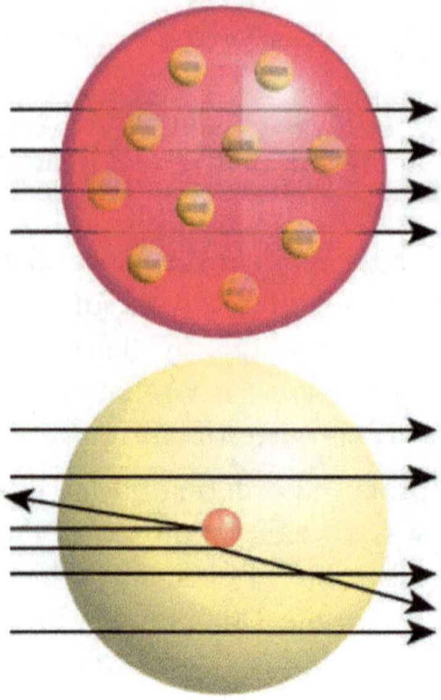

Top: Expected results: alpha particles passing through the plum pudding model of the atom undisturbed.

Bottom: Observed results: a small portion of the particles were deflected, indicating a small, concentrated charge.

At the turn of the twentieth century the theoretical underpinnings of chemistry were finally understood due to a series of remarkable discoveries that succeeded in probing and discovering the very nature of the internal structure of atoms. In 1897, J. J. Thomson of Cambridge University discovered the electron and soon after the French scientist Becquerel as well as the couple Pierre and Marie Curie investigated the phenomenon of radioactivity. In a series of pioneering scattering experiments Ernest Rutherford at the University of Manchester discovered the internal structure of the atom and the existence of the proton, classified and explained the different types of radioactivity and successfully transmuted the first element by bombarding nitrogen with alpha particles.

His work on atomic structure was improved on by his students, the Danish physicist Niels Bohr and Henry Moseley. The electronic theory of chemical bonds and molecular orbitals was developed by the American scientists Linus Pauling and Gilbert N. Lewis.

The year 2011 was declared by the United Nations as the International Year of Chemistry.. It was an initiative of the International Union of Pure and Applied Chemistry, and of the United Nations Educational, Scientific, and Cultural Organization and involves chemical societies, academics, and institutions worldwide and relied on individual initiatives to organize local and regional activities.

Principles of Modern Chemistry

The current model of atomic structure is the quantum mechanical model.. Traditional chemistry starts with the study of elementary particles, atoms, molecules,. substances, metals, crystals and other aggregates of matter. This matter can be studied in solid, liquid, or gas states, in isolation or in combination. The interactions, reactions and transformations that are studied in chemistry are usually the result of interactions between atoms, leading to rearrangements of the chemical bonds which hold atoms together. Such behaviors are studied in a chemistry laboratory.

The chemistry laboratory stereotypically uses various forms of laboratory glassware. However glassware is not central to chemistry, and a great deal of experimental (as well as applied/industrial) chemistry is done without it.

A chemical reaction is a transformation of some substances into one or more different substances.. The basis of such a chemical transformation is the rearrangement of electrons in the chemical bonds between atoms. It can be symbolically depicted through a chemical equation, which usually involves atoms as subjects. The number of atoms on the left and the right in the equation for a chemical transformation is equal. (When the number of atoms on either side is unequal, the transformation is referred to as a nuclear reaction or radioactive decay.) The type of chemical reactions a substance may undergo and the energy changes that may accompany it are constrained by certain basic rules, known as chemical laws.

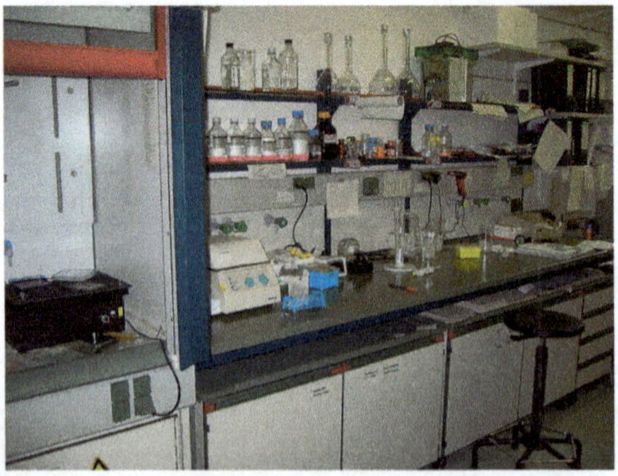

Laboratory, Institute of Biochemistry, University of Cologne in Germany.

Energy and entropy considerations are invariably important in almost all chemical studies. Chemical substances are classified in terms of their structure, phase, as well as their chemical compositions. They can be analyzed using the tools of chemical analysis, e.g. spectroscopy and chromatography. Scientists engaged in chemical research are known as chemists.. Most chemists specialize in one or more sub-disciplines. Several concepts are essential for the study of chemistry; some of them are:.

Matter

In chemistry, matter is defined as anything that has rest mass and volume (it takes up space) and is made up of particles. The particles that make up matter have rest mass as well - not all particles have rest mass, such as the photon. Matter can be a pure chemical substance or a mixture of substances..

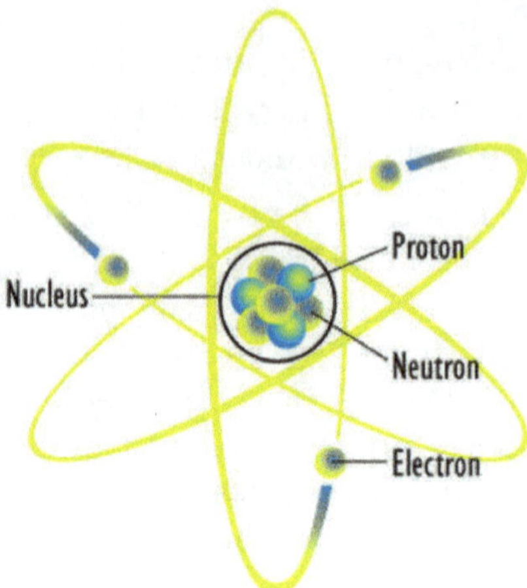

A diagram of an atom based on the Rutherford model

The atom is the basic unit of chemistry. It consists of a dense core called the atomic nucleus surrounded by a space called the electron cloud. The nucleus is made up of positively charged protons and uncharged neutrons (together called nucleons), while the electron cloud consists of negatively charged electrons which orbit the nucleus. In a neutral atom, the negatively charged electrons balance out the positive charge of the protons. The nucleus is dense; the mass of a nucleon is 1,836 times that of an electron, yet the radius of an atom is about 10,000 times that of its nucleus...

The atom is also the smallest entity that can be envisaged to retain the chemical properties of the element, such as electronegativity, ionization potential, preferred oxidation state(s), coordination number, and preferred types of bonds to form (e.g., metallic, ionic, covalent).

Element

A chemical element is a pure substance which is composed of a single type of atom, characterized by its particular number of protons in the nuclei of its atoms, known as the atomic number and represented by the symbol Z. The mass number is the sum of the number of protons and neutrons in a nucleus. Although all the nuclei of all atoms belonging to one element will have the same atomic number, they may not necessarily have the same mass number; atoms of an element which have different mass numbers are known as isotopes. For example, all atoms with 6 protons in their nuclei are atoms of the chemical element carbon, but atoms of carbon may have mass numbers of 12 or 13..

Group→ ↓Period	1	2	3	4	5	6	7	8	9	10	11	12	13	14	15	16	17	18
1	1 H																	2 He
2	3 Li	4 Be											5 B	6 C	7 N	8 O	9 F	10 Ne
3	11 Na	12 Mg											13 Al	14 Si	15 P	16 S	17 Cl	18 Ar
4	19 K	20 Ca	21 Sc	22 Ti	23 V	24 Cr	25 Mn	26 Fe	27 Co	28 Ni	29 Cu	30 Zn	31 Ga	32 Ge	33 As	34 Se	35 Br	36 Kr
5	37 Rb	38 Sr	39 Y	40 Zr	41 Nb	42 Mo	43 Tc	44 Ru	45 Rh	46 Pd	47 Ag	48 Cd	49 In	50 Sn	51 Sb	52 Te	53 I	54 Xe
6	55 Cs	56 Ba	*	72 Hf	73 Ta	74 W	75 Re	76 Os	77 Ir	78 Pt	79 Au	80 Hg	81 Tl	82 Pb	83 Bi	84 Po	85 At	86 Rn
7	87 Fr	88 Ra	**	104 Rf	105 Db	106 Sg	107 Bh	108 Hs	109 Mt	110 Ds	111 Rg	112 Cn	113 Uut	114 Fl	115 Uup	116 Lv	117 Uus	118 Uuo

*	57 La	58 Ce	59 Pr	60 Nd	61 Pm	62 Sm	63 Eu	64 Gd	65 Tb	66 Dy	67 Ho	68 Er	69 Tm	70 Yb	71 Lu
**	89 Ac	90 Th	91 Pa	92 U	93 Np	94 Pu	95 Am	96 Cm	97 Bk	98 Cf	99 Es	100 Fm	101 Md	102 No	103 Lr

Standard form of the periodic table of chemical elements. The colors represent different categories of elements

The standard presentation of the chemical elements is in the periodic table, which orders elements by atomic number. The periodic table is arranged in groups, or columns, and periods, or rows. The periodic table is useful in identifying periodic trends..

Compound

Carbon dioxide (CO_2), an example of a chemical compound

A compound is a pure chemical substance composed of more than one element. The properties of a compound bear little similarity to those of its elements.. The standard nomenclature of compounds is set by the International Union of Pure and Applied Chemistry (IUPAC). Organic compounds are named according to the organic nomenclature system.. Inorganic compounds are named according to the inorganic nomenclature system.. In addition the Chemical Abstracts Service has devised a method to index chemical substances. In this scheme each chemical substance is identifiable by a number known as its CAS registry number.

Molecule

A molecule is the smallest indivisible portion of a pure chemical substance that has its unique set of chemical properties, that is, its potential to undergo a certain set of chemical reactions with other substances. However, this definition only works well for substances that are composed of molecules, which is not true of many substances (see below). Molecules are typically a set of atoms bound together by covalent bonds, such that the structure is electrically neutral and all valence electrons are paired with other electrons either in bonds or in lone pairs.

A ball-and-stick representation of the caffeine molecule ($C_8H_10N_4O_2$).

Thus, molecules exist as electrically neutral units, unlike ions. When this rule is broken, giving the "molecule" a charge, the result is sometimes named a molecular ion or a polyatomic ion. However, the discrete and separate nature of the molecular concept usually requires that molecular ions be present only in well-separated form, such as a directed beam in a vacuum in a mass spectrometer. Charged polyatomic collections residing in solids (for example, common sulfate or nitrate ions) are generally not considered "molecules" in chemistry.

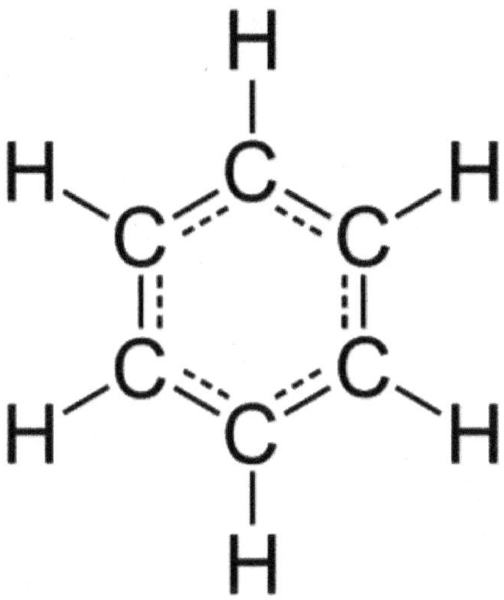

A 2-D skeletal model of a benzene molecule (C_6H_6)

The "inert" or noble gas elements (helium, neon, argon, krypton, xenon and radon) are composed of lone atoms as their smallest discrete unit, but the other isolated chemical elements consist of either molecules or networks of atoms bonded to each other in some way. Identifiable molecules compose familiar substances such as water, air, and many organic compounds like alcohol, sugar, gasoline, and the various pharmaceuticals.

However, not all substances or chemical compounds consist of discrete molecules, and indeed most of the solid substances that make up the solid crust, mantle, and core of the Earth are chemical compounds without molecules. These other types of substances, such as ionic compounds and network solids, are organized in such a way as to lack the existence of identifiable molecules per se. Instead, these substances are discussed in terms of formula units or unit cells as the smallest repeating structure within the substance. Examples of such substances are mineral salts (such as table salt), solids like carbon and diamond, metals, and familiar silica and silicate minerals such as quartz and granite.

One of the main characteristics of a molecule is its geometry often called its structure. While the structure of diatomic, triatomic or tetra atomic molecules may be trivial, (linear, angular pyramidal etc.) the structure of polyatomic molecules, that are constituted of more than six atoms (of several elements) can be crucial for its chemical nature.

Substance and Mixture

Examples of pure chemical substances. From left to right: the elements tin (Sn) and sulfur (S), diamond (an allotrope of carbon), sucrose (pure sugar), and sodium chloride (salt) and sodium bicarbonate (baking soda), which are both ionic compounds.

A chemical substance is a kind of matter with a definite composition and set of properties.. A collection of substances is called a mixture. Examples of mixtures are air and alloys..

Mole and Amount of Substance

The mole is a unit of measurement that denotes an amount of substance (also called chemical amount). The mole is defined as the number of atoms found in exactly 0.012 kilogram (or 12 grams) of carbon-12, where the carbon-12 atoms are unbound, at rest and in their ground state.. The number of entities per mole is known as the Avogadro constant, and is determined empirically to be approximately 6.022×10^{23} mol^{-1}.. Molar concentration is the amount of a particular substance per volume of solution, and is commonly reported in $moldm^{-3}$..

Phase

In addition to the specific chemical properties that distinguish different chemical classifications, chemicals can exist in several phases. For the most part, the chemical classifications are independent of these bulk phase classifications; however, some more exotic phases are incompatible with certain chemical properties. A phase is a set of states of a chemical system that have similar bulk structural properties, over a range of conditions, such as pressure or temperature.

Physical properties, such as density and refractive index tend to fall within values characteristic of the phase. The phase of matter is defined by the phase transition, which is when energy put into or taken out of the system goes into rearranging the structure of the system, instead of changing the bulk conditions.

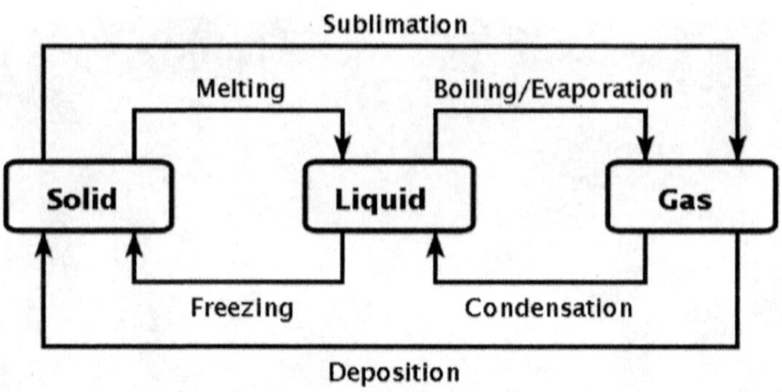

Example of phase changes

Sometimes the distinction between phases can be continuous instead of having a discrete boundary, in this case the matter is considered to be in a supercritical state. When three states meet based on the conditions, it is known as a triple point and since this is invariant, it is a convenient way to define a set of conditions.

The most familiar examples of phases are solids, liquids, and gases. Many substances exhibit multiple solid phases. For example, there are three phases of solid iron (alpha, gamma, and delta) that vary based on temperature and pressure. A principal difference between solid phases is the crystal structure, or arrangement, of the atoms. Another phase commonly encountered in the study of chemistry is the aqueous phase, which is the state of substances dissolved in aqueous solution (that is, in water).

Less familiar phases include plasmas, Bose–Einstein condensates and fermionic condensates and the paramagnetic and ferromagnetic phases of magnetic materials. While most familiar phases deal with three-dimensional systems, it is also possible to define analogs in two-dimensional systems, which has received attention for its relevance to systems in biology.

Bonding

Atoms sticking together in molecules or crystals are said to be bonded with one another. A chemical bond may be visualized as the multipole balance between the positive charges in the nuclei and the negative charges oscillating about them.. More than simple attraction and repulsion, the energies and distributions characterize the availability of an electron to bond to another atom.

A chemical bond can be a covalent bond, an ionic bond, a hydrogen bond or just because of Van der Waals force. Each of these kinds of bonds is ascribed to some potential. These potentials create the interactions which hold atoms together in molecules or crystals. In many simple compounds, valence bond theory, the Valence Shell Electron Pair Repulsion model (VSEPR), and the concept of oxidation number can be used to explain molecular structure and composition.

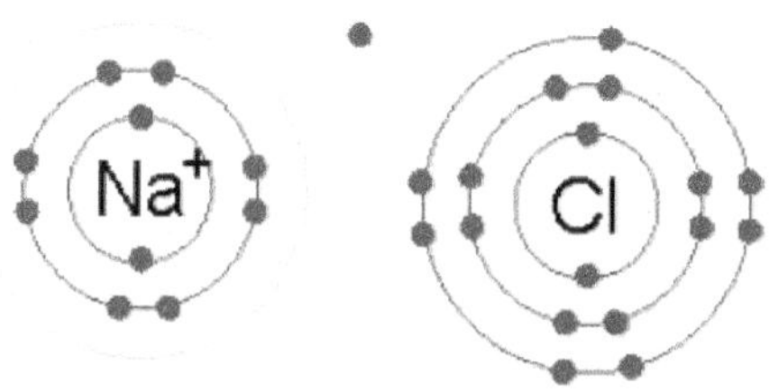

An animation of the process of ionic bonding between sodium (Na) and chlorine (Cl) to form sodium chloride, or common table salt. Ionic bonding involves one atom taking valence electrons from another (as opposed to sharing, which occurs in covalent bonding)

An ionic bond is formed when a metal loses one or more of its electrons, becoming a positively charged cation, and the electrons are then gained by the non-metal atom, becoming a negatively charged anion. The two oppositely charged ions attract one another, and the ionic bond is the electrostatic force of attraction between them. For example, sodium (Na), a metal, loses one electron to become an Na+ cation while chlorine (Cl), a non-metal, gains this electron to become Cl−. The ions are held together due to electrostatic attraction, and that compound sodium chloride (NaCl), or common table salt, is formed.

In a covalent bond, one or more pairs of valence electrons are shared by two atoms: the resulting electrically neutral group of bonded atoms is termed a molecule. Atoms will share valence electrons in such a way as to create a noble gas electron configuration (eight electrons in their outermost shell) for each atom. Atoms that tend to combine in such a way that they each have eight electrons in their valence shell are said to follow the octet rule. However, some elements like hydrogen and lithium need only two electrons in their outermost shell to attain this stable configuration; these atoms are said to follow the duet rule, and in this way they are reaching the electron configuration of the noble gas helium, which has two electrons in its outer shell.

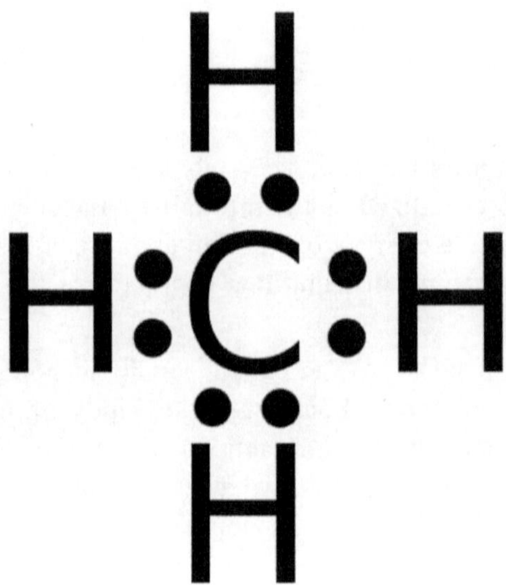

In the methane molecule (CH_4), the carbon atom shares a pair of valence electrons with each of the four hydrogen atoms. Thus, the octet rule is satisfied for C-atom (it has eight electrons in its valence shell) and the duet rule is satisfied for the H-atoms (they have two electrons in their valence shells).

Similarly, theories from classical physics can be used to predict many ionic structures. With more complicated compounds, such as metal complexes, valence bond theory is less applicable and alternative approaches, such as the molecular orbital theory, are generally used. See diagram on electronic orbitals.

Energy

In the context of chemistry, energy is an attribute of a substance as a consequence of its atomic, molecular or aggregate structure. Since a chemical transformation is accompanied by a change in one or more of these kinds of structures, it is invariably accompanied by an increase or decrease of energy of the substances involved. Some energy is transferred between the surroundings and the reactants of the reaction in the form of heat or light; thus the products of a reaction may have more or less energy than the reactants.

A reaction is said to be exergonic if the final state is lower on the energy scale than the initial state; in the case of endergonic reactions the situation is the reverse. A reaction is said to be exothermic if the reaction releases heat to the surroundings; in the case of endothermic reactions, the reaction absorbs heat from the surroundings.

Chemical reactions are invariably not possible unless the reactants surmount an energy barrier known as the activation energy. The speed of a chemical reaction (at given temperature T) is related to the activation energy E, by the Boltzmann's population factor $e^{-E/kT}$ that is the probability of a molecule to have energy greater than or equal to E at the given temperature T. This exponential dependence of a reaction rate on temperature is known as the Arrhenius equation. The activation energy necessary for a chemical reaction to occur can be in the form of heat, light, electricity or mechanical force in the form of ultrasound..

A related concept free energy, which also incorporates entropy considerations, is a very useful means for predicting the feasibility of a reaction and determining the state of equilibrium of a chemical reaction, in chemical thermodynamics. A reaction is feasible only if the total change in the Gibbs free energy is negative, $\Delta G \leq 0$ if it is equal to zero the chemical reaction is said to be at equilibrium.

There exist only limited possible states of energy for electrons, atoms and molecules. These are determined by the rules of quantum mechanics, which require quantization of energy of a bound system. The atoms/molecules in a higher energy state are said to be excited. The molecules/atoms of substance in an excited energy state are often much more reactive; that is, more amenable to chemical reactions.

The phase of a substance is invariably determined by its energy and the energy of its surroundings. When the intermolecular forces of a substance are such that the energy of the surroundings is not sufficient to overcome them, it occurs in a more ordered phase like liquid or solid as is the case with water (H_2O); a liquid at room temperature because its molecules are bound by hydrogen bonds.. Whereas hydrogen sulfide (H_2S) is a gas at room temperature and standard pressure, as its molecules are bound by weaker dipole-dipole interactions.

The transfer of energy from one chemical substance to another depends on the size of energy quanta emitted from one substance. However, heat energy is often transferred more easily from almost any substance to another because the phonons responsible for vibrational and rotational energy levels in a substance have much less energy than photons invoked for the electronic energy transfer. Thus, because vibrational and rotational energy levels are more closely spaced than electronic energy levels, heat is more easily transferred between substances relative to light or other forms of electronic energy. For example, ultraviolet electromagnetic radiation is not transferred with as much efficacy from one substance to another as thermal or electrical energy.

The existence of characteristic energy levels for different chemical substances is useful for their identification by the analysis of spectral lines. Different kinds of spectra are often used in chemical spectroscopy, e.g. IR, microwave, NMR, ESR, etc. Spectroscopy is also used to identify the composition of remote objects - like stars and distant galaxies - by analyzing their radiation spectra.

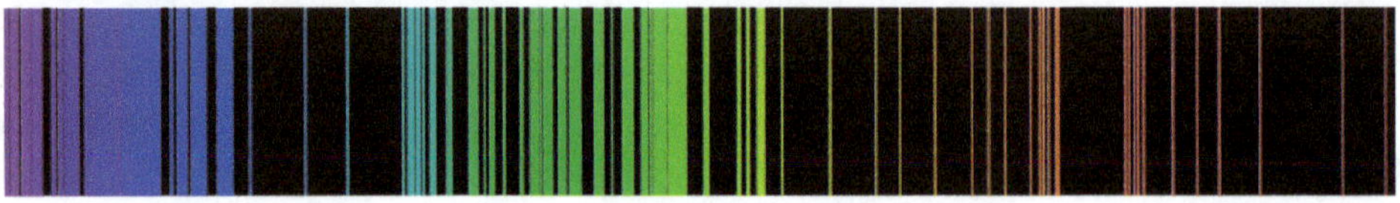

Emission spectrum of iron

The term chemical energy is often used to indicate the potential of a chemical substance to undergo a transformation through a chemical reaction or to transform other chemical substances.

Reaction

When a chemical substance is transformed as a result of its interaction with another substance or with energy, a chemical reaction is said to have occurred. A chemical reaction is therefore a concept related to the "reaction" of a substance when it comes in close contact with another, whether as a mixture or a solution; exposure to some form of energy, or both. It results in some energy exchange between the constituents of the reaction as well as with the system environment, which may be designed vessels—often laboratory glassware.

Chemical reactions can result in the formation or dissociation of molecules, that is, molecules breaking apart to form two or smaller molecules, or rearrangement of atoms within or across molecules. Chemical reactions usually involve the making or breaking of chemical bonds. Oxidation, reduction, dissociation, acid-base neutralization and molecular rearrangement are some of the commonly used kinds of chemical reactions.

During chemical reactions, bonds between atoms break and form, resulting in different substances with different properties. In a blast furnace, iron oxide, a compound, reacts with carbon monoxide to form iron, one of the chemical elements, and carbon dioxide.

A chemical reaction can be symbolically depicted through a chemical equation. While in a non-nuclear chemical reaction the number and kind of atoms on both sides of the equation are equal, for a nuclear reaction this holds true only for the nuclear particles viz. protons and neutrons..

The sequence of steps in which the reorganization of chemical bonds may be taking place in the course of a chemical reaction is called its mechanism. A chemical reaction can be envisioned to take place in a number of steps, each of which may have a different speed. Many reaction intermediates with variable stability can thus be envisaged during the course of a reaction. Reaction mechanisms are proposed to explain the kinetics and the relative product mix of a reaction. Many physical chemists specialize in exploring and proposing the mechanisms of various chemical reactions. Several empirical rules, like the Woodward–Hoffmann rules often come in handy while proposing a mechanism for a chemical reaction.

According to the IUPAC gold book, a chemical reaction is "a process that results in the interconversion of chemical species.". Accordingly, a chemical reaction may be an elementary reaction or a stepwise reaction. An additional caveat is made, in that this definition includes cases where the interconversion of conformers is experimentally observable. Such detectable chemical reactions normally involve sets of molecular entities as indicated by this definition, but it is often conceptually convenient to use the term also for changes involving single molecular entities (i.e. 'microscopic chemical events').

Ions and Salts

An ion is a charged species, an atom or a molecule, that has lost or gained one or more electrons. When an atom loses an electron and thus has more protons than electrons, the atom is a positively charged ion or cation. When an atom gains an electron and thus has more electrons than protons, the atom is a negatively charged ion or anion. Cations and anions can form a crystalline lattice of neutral salts, such as the Na^+ and Cl^- ions forming sodium chloride, or NaCl. Examples of polyatomic ions that do not split up during acid-base reactions are hydroxide (OH^-) and phosphate (PO_4^{3-}).

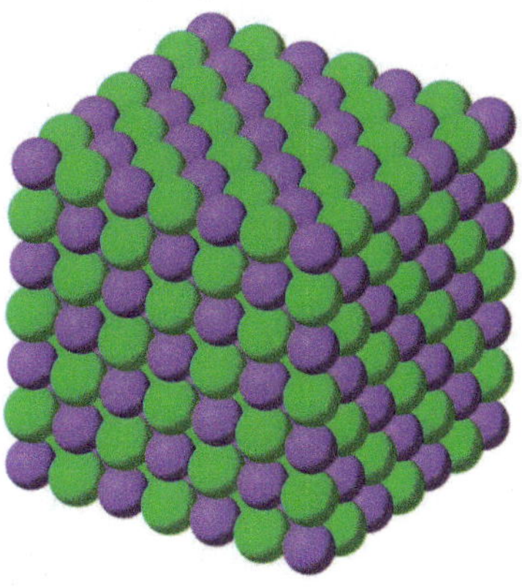

The crystal lattice structure of potassium chloride (KCl), a salt which is formed due to the attraction of K^+ cations and Cl^- anions. Note how the overall charge of the ionic compound is zero.

Plasma is composed of gaseous matter that has been completely ionized, usually through high temperature.

Acidity and Basicity

When hydrogen bromide (HBr), pictured, is dissolved in water, it forms the strong acid hydrobromic acid

A substance can often be classified as an acid or a base. There are several different theories which explain acid-base behavior. The simplest is Arrhenius theory, which states than an acid is a substance that produces hydronium ions when it is dissolved in water, and a base is one that produces hydroxide ions when dissolved in water. According to Brønsted–Lowry acid–base theory, acids are substances that donate a positive hydrogen ion to another substance in a chemical reaction; by extension, a base is the substance which receives that hydrogen ion.

A third common theory is Lewis acid-base theory, which is based on the formation of new chemical bonds. Lewis theory explains that an acid is a substance which is capable of accepting a pair of electrons from another substance during the process of bond formation, while a base is a substance which can provide a pair of electrons to form a new bond. According to this theory, the crucial things being exchanged are charges.. There are several other ways in which a substance may be classified as an acid or a base, as is evident in the history of this concept..

Acid strength is commonly measured by two methods. One measurement, based on the Arrhenius definition of acidity, is pH, which is a measurement of the hydronium ion concentration in a solution, as expressed on a negative logarithmic scale. Thus, solutions that have a low pH have a high hydronium ion concentration, and can be said to be more acidic. The other measurement, based on the Brønsted–Lowry definition, is the acid dissociation constant (K_a), which measures the relative ability of a substance to act as an acid under the Brønsted–Lowry definition of an acid. That is, substances with a higher K_a are more likely to donate hydrogen ions in chemical reactions than those with lower K_a values.

Redox

Redox (reduction-oxidation) reactions include all chemical reactions in which atoms have their oxidation state changed by either gaining electrons (reduction) or losing electrons (oxidation). Substances that have the ability to oxidize other substances are said to be oxidative and are known as oxidizing agents, oxidants or oxidizers. An oxidant removes electrons from another substance. Similarly, substances that have the ability to reduce other substances are said to be reductive and are known as reducing agents, reductants, or reducers.

A reductant transfers electrons to another substance, and is thus oxidized itself. And because it "donates" electrons it is also called an electron donor. Oxidation and reduction properly refer to a change in oxidation number—the actual transfer of electrons may never occur. Thus, oxidation is better defined as an increase in oxidation number, and reduction as a decrease in oxidation number.

Equilibrium

Although the concept of equilibrium is widely used across sciences, in the context of chemistry, it arises whenever a number of different states of the chemical composition are possible, as for example, in a mixture of several chemical compounds that can react with one another, or when a substance can be present in more than one kind of phase.

A system of chemical substances at equilibrium, even though having an unchanging composition, is most often not static; molecules of the substances continue to react with one another thus giving rise to a dynamic equilibrium. Thus the concept describes the state in which the parameters such as chemical composition remain unchanged over time.

Chemical Laws

Chemical reactions are governed by certain laws, which have become fundamental concepts in chemistry. Some of them are:

- Avogadro's law
- Beer–Lambert law
- Boyle's law (1662, relating pressure and volume)
- Charles's law (1787, relating volume and temperature)
- Fick's laws of diffusion
- Gay-Lussac's law (1809, relating pressure and temperature)
- Le Chatelier's principle
- Henry's law

- Hess's law
- Law of conservation of energy leads to the important concepts of equilibrium, thermodynamics, and kinetics.
- Law of conservation of mass continues to be conserved in isolated systems, even in modern physics. However, special relativity shows that due to mass–energy equivalence, whenever non-material "energy" (heat, light, kinetic energy) is removed from a non-isolated system, some mass will be lost with it. High energy losses result in loss of weighable amounts of mass, an important topic in nuclear chemistry.
- Law of definite composition, although in many systems (notably biomacromolecules and minerals) the ratios tend to require large numbers, and are frequently represented as a fraction.
- Law of multiple proportions
- Raoult's law

PRACTICE

Subdisciplines

Chemistry is typically divided into several major sub-disciplines. There are also several main cross-disciplinary and more specialized fields of chemistry..

- Analytical chemistry is the analysis of material samples to gain an understanding of their chemical composition and structure. Analytical chemistry incorporates standardized experimental methods in chemistry. These methods may be used in all subdisciplines of chemistry, excluding purely theoretical chemistry.
- Biochemistry is the study of the chemicals, chemical reactions and chemical interactions that take place in living organisms. Biochemistry and organic chemistry are closely related, as in medicinal chemistry or neurochemistry. Biochemistry is also associated with molecular biology and genetics.
- Inorganic chemistry is the study of the properties and reactions of inorganic compounds. The distinction between organic and inorganic disciplines is not absolute and there is much overlap, most importantly in the sub-discipline of organometallic chemistry.
- Materials chemistry is the preparation, characterization, and understanding of substances with a useful function. The field is a new breadth of study in graduate programs, and it integrates elements from all classical areas of chemistry with a focus on fundamental issues that are unique to materials. Primary systems of study include the chemistry of condensed phases (solids, liquids, polymers) and interfaces between different phases.
- Neurochemistry is the study of neurochemicals; including transmitters, peptides, proteins, lipids, sugars, and nucleic acids; their interactions, and the roles they play in forming, maintaining, and modifying the nervous system.

- Nuclear chemistry is the study of how subatomic particles come together and make nuclei. Modern Transmutation is a large component of nuclear chemistry, and the table of nuclides is an important result and tool for this field.
- Organic chemistry is the study of the structure, properties, composition, mechanisms, and reactions of organic compounds. An organic compound is defined as any compound based on a carbon skeleton.
- Physical chemistry is the study of the physical and fundamental basis of chemical systems and processes. In particular, the energetics and dynamics of such systems and processes are of interest to physical chemists. Important areas of study include chemical thermodynamics, chemical kinetics, electrochemistry, statistical mechanics, spectroscopy, and more recently, astrochemistry.. Physical chemistry has large overlap with molecular physics. Physical chemistry involves the use of infinitesimal calculus in deriving equations. It is usually associated with quantum chemistry and theoretical chemistry. Physical chemistry is a distinct discipline from chemical physics, but again, there is very strong overlap.
- Theoretical chemistry is the study of chemistry via fundamental theoretical reasoning (usually within mathematics or physics). In particular the application of quantum mechanics to chemistry is called quantum chemistry. Since the end of the Second World War, the development of computers has allowed a systematic development of computational chemistry, which is the art of developing and applying computer programs for solving chemical problems. Theoretical chemistry has large overlap with (theoretical and experimental) condensed matter physics and molecular physics.

Other disciplines within chemistry are traditionally grouped by the type of matter being studied or the kind of study. These include inorganic chemistry, the study of inorganic matter; organic chemistry, the study of organic (carbon-based) matter; biochemistry, the study of substances found in biological organisms; physical chemistry, the study of chemical processes using physical concepts such as thermodynamics and quantum mechanics; and analytical chemistry, the analysis of material samples to gain an understanding of their chemical composition and structure. Many more specialized disciplines have emerged in recent years, e.g. neurochemistry the chemical study of the nervous system (see subdisciplines).

Other fields include agrochemistry, astrochemistry (and cosmochemistry), atmospheric chemistry, chemical engineering, chemical biology, chemo-informatics, electrochemistry, environmental chemistry, femtochemistry, flavor chemistry, flow chemistry, geochemistry, green chemistry, histochemistry, history of chemistry, hydrogenation chemistry, immunochemistry, marine chemistry, materials science, mathematical chemistry, mechanochemistry, medicinal chemistry, molecular biology, molecular mechanics, nanotechnology, natural product chemistry, oenology, organometallic chemistry, petrochemistry, pharmacology, photochemistry, physical organic chemistry, phytochemistry, polymer chemistry, radiochemistry, solid-state chemistry, sonochemistry, supramolecular chemistry, surface chemistry, synthetic chemistry, thermochemistry, and many others.

Chemical Industry

The chemical industry represents an important economic activity worldwide. The global top 50 chemical producers in 2013 had sales of US$980.5 billion with a profit margin of 10.3%..

REFERENCES

1. "What is Chemistry?". Chemweb.ucc.ie. Retrieved 2011-06-12.
2. Chemistry. (n.d.). Merriam-Webster's Medical Dictionary. Retrieved August 19, 2007.
3. Theodore L. Brown, H. Eugene Lemay, Bruce Edward Bursten, H. Lemay. Chemistry: The Central Science. Prentice Hall; 8 edition (1999). ISBN 0-13-010310-1. Pages 3–4.
4. Chemistry occupies an intermediate position in a hierarchy of the sciences by reductive level between physics and biology. Carsten Reinhardt. Chemical Sciences in the 20th Century: Bridging Boundaries. Wiley-VCH, 2001. ISBN 3-527-30271-9. Pages 1–2.
5. Bunge, M. (1982). "Is chemistry a branch of physics?". Journal for the General Philosophy of Science - Zeitschrift für allgemeine Wissenschaftstheorie. 13 (2): 209–223. doi:10.1007/BF01801556.
6. "History of Alchemy". Alchemy Lab. Retrieved 2011-06-12.
7. Strathern, P. (2000). Mendeleyev's Dream – the Quest for the Elements. New York: Berkley Books.
8. Alchemy", entry in The Oxford English Dictionary, J. A. Simpson and E. S. C. Weiner, vol. 1, 2nd ed., 1989, ISBN 0-19-861213-3.
9. p. 854, "Arabic alchemy", Georges C. Anawati, pp. 853–885 in Encyclopedia of the history of Arabic science, eds. Roshdi Rashed and Régis Morelon, London: Routledge, 1996, vol. 3, ISBN 0-415-12412-3.
10. Weekley, Ernest (1967). Etymological Dictionary of Modern English. New York: Dover Publications. ISBN 0-486-21873-2
11. Boyle, Robert (1661). The Sceptical Chymist. New York: Dover Publications, Inc. (reprint). ISBN 0-486-42825-7.
12. Glaser, Christopher (1663). Traite de la chymie. Paris. as found in: Kim, Mi Gyung (2003). Affinity, That Elusive Dream - A Genealogy of the Chemical Revolution. The MIT Press. ISBN 0-262-11273-6.
13. Stahl, George, E. (1730). Philosophical Principles of Universal Chemistry. London.
14. Dumas, J. B. (1837). 'Affinite' (lecture notes), vii, pg 4. "Statique chimique", Paris: Academie des Sciences
15. Pauling, Linus (1947). General Chemistry. Dover Publications, Inc. ISBN 0-486-65622-5.

16. Chang, Raymond (1998). Chemistry, 6th Ed. New York: McGraw Hill. ISBN 0-07-115221-0.
17. First chemists, February 13, 1999, New Scientist
18. Barnes, Ruth. Textiles in Indian Ocean Societies. Routledge. p. 1.
19. Lucretius (50 BCE). "de Rerum Natura (On the Nature of Things)". The Internet Classics Archive. Massachusetts Institute of Technology. Retrieved 9 January 2007. Check date values in: |date= (help)
20. Simpson, David (29 June 2005). "Lucretius (c. 99 - c. 55 BCE)". The Internet History of Philosophy. Retrieved 2007-01-09.
21. Strodach, George K. (2012). The Art of Happiness. New York: Penguin Classics. pp. 7–8. ISBN 0-14-310721-6.
22. "International Year of Chemistry - The History of Chemistry". G.I.T. Laboratory Journal Europe. Feb 25, 2011. Retrieved March 12, 2013.
23. Bryan H. Bunch & Alexander Hellemans (2004). The History of Science and Technology. Houghton Mifflin Harcourt. p. 88. ISBN 0-618-22123-9.
24. Morris Kline (1985) Mathematics for the nonmathematician. Courier Dover Publications. p. 284. ISBN 0-486-24823-2
25. Marcelin Berthelot, Collection des anciens alchimistes grecs (3 vol., Paris, 1887–1888, p.161); F. Sherwood Taylor, "The Origins of Greek Alchemy," Ambix 1 (1937), 40.
26. Marmura, Michael E.; Nasr, Seyyed Hossein (1965). "An Introduction to Islamic Cosmological Doctrines. Conceptions of Nature and Methods Used for Its Study by the Ikhwan Al-Safa'an, Al-Biruni, and Ibn Sina by Seyyed Hossein Nasr". Speculum. 40 (4): 744–746. doi:10.2307/2851429. JSTOR 2851429.
27. Robert Briffault (1938). The Making of Humanity, p. 196-197.
28. Alakbarov, Farid (2001). "A 13th-Century Darwin? Tusi's Views on Evolution". Azerbaijan International. 9: 2.
29. Derewenda, Zygmunt S.; Derewenda, ZS (2007). "On wine, chirality and crystallography". Acta Crystallographica Section A. 64 (Pt 1): 246–258 .. Bibcode:2008AcCrA..64..246D. doi:10.1107/S0108767307054293. PMID 18156689.
30. John Warren (2005). "War and the Cultural Heritage of Iraq: a sadly mismanaged affair", Third World Quarterly, Volume 26, Issue 4 & 5, p. 815-830.
31. Dr. A. Zahoor (1997), JABIR IBN HAIYAN (Geber)
32. Paul Vallely, How Islamic inventors changed the world, The Independent, 10 March 2006
33. Kraus, Paul, Jâbir ibn Hayyân, Contribution à l'histoire des idées scientifiques dans l'Islam. I. Le corpus des écrits jâbiriens. II. Jâbir et la science grecque,. Cairo (1942-1943). Repr. By Fuat Sezgin, (Natural Sciences in Islam. 67-68), Frankfurt. 2002: "To form an idea of the historical place of Jabir›s alchemy and to tackle the problem of its sources, it is advisable to compare it with what remains

to us of the alchemical literature in the Greek language. One knows in which miserable state this literature reached us. Collected by Byzantine scientists from the tenth century, the corpus of the Greek alchemists is a cluster of incoherent fragments, going back to all the times since the third century until the end of the Middle Ages." "The efforts of Berthelot and Ruelle to put a little order in this mass of literature led only to poor results, and the later researchers, among them in particular Mrs. Hammer-Jensen, Tannery, Lagercrantz, von Lippmann, Reitzenstein, Ruska, Bidez, Festugiere and others, could make clear only few points of detail...The study of the Greek alchemists is not very encouraging. An even surface examination of the Greek texts shows that a very small part only was organized according to true experiments of laboratory: even the supposedly technical writings, in the state where we find them today, are unintelligible nonsense which refuses any interpretation.It is different with Jabir›s alchemy. The relatively clear description of the processes and the alchemical apparatuses, the methodical classification of the substances, mark an experimental spirit which is extremely far away from the weird and odd esotericism of the Greek texts. The theory on which Jabir supports his operations is one of clearness and of an impressive unity. More than with the other Arab authors, one notes with him a balance between theoretical teaching and practical teaching, between the `ilm and the `amal. In vain one would seek in the Greek texts a work as systematic as that which is presented for example in the Book of Seventy."

34. "Robert Boyle, Founder of Modern Chemistry" Harry Sootin (2011)
35. "History - Robert Boyle (1627–1691)". BBC. Retrieved 2011-06-12.
36. Eagle, Cassandra T.; Jennifer Sloan (1998). "Marie Anne Paulze Lavoisier: The Mother of Modern Chemistry". The Chemical Educator. 3 (5): 1–18. doi:10.1007/s00897980249a.
37. Mi Gyung Kim (2003). Affinity, that Elusive Dream: A Genealogy of the Chemical Revolution. MIT Press. p. 440. ISBN 0-262-11273-6.
38. Chemistry 412 course notes. "A Brief History of the Development of Periodic Table". Western Oregon University. Retrieved July 20, 2015.
39. It is surely true that had Mendeleev never lived modern chemists would be using a Periodic Table" and "Dmitri Mendeleev". Royal Society of Chemistry. Retrieved July 18, 2015.
40. Davy, Humphry (1808). "On some new Phenomena of Chemical Changes produced by Electricity, particularly the Decomposition of the fixed Alkalies, and the Exhibition of the new Substances, which constitute their Bases". Philosophical Transactions of the Royal Society. Royal Society of London. 98 (0): 1–45. doi:10.1098/rstl.1808.0001.
41. Winter, Mark. "WebElements: the periodic table on the web". The University of Sheffield. Archived from the original on January 4, 2014. Retrieved January 27, 2014.
42. "Julius Lothar Meyer and Dmitri Ivanovich Mendeleev". Chemical Heritage Foundation. Retrieved July 18, 2015.

43. "What makes these family likenesses among the elements? In the 1860s everyone was scratching their heads about that, and several scientists moved towards rather similar answers. The man who solved the problem most triumphantly was a young Russian called Dmitri Ivanovich Mendeleev, who visited the salt mine at Wieliczka in 1859." Bronowski, Jacob (1973). The Ascent of Man. Little, Brown and Company. p. 322. ISBN 0-316-10930-4.
44. Ihde, Aaron John (1984). The Development of Modern Chemistry. Courier Dover Publications. p. 164. ISBN 0-486-64235-6.
45. "Chemistry". Chemistry2011.org. Retrieved 2012-03-10.
46. "chemical bonding". Britannica. Encyclopædia Britannica. Retrieved 1 November 2012.
47. Matter: Atoms from Democritus to Dalton by Anthony Carpi, Ph.D.
48. IUPAC Gold Book Definition
49. "California Occupational Guide Number 22: Chemists". Calmis.ca.gov. 1999-10-29. Retrieved 2011-06-12.
50. "General Chemistry Online - Companion Notes: Matter". Antoine.frostburg.edu. Retrieved 2011-06-12.
51. Armstrong, James (2012). General, Organic, and Biochemistry: An Applied Approach. Brooks/Cole. p. 48. ISBN 978-0-534-49349-3.
52. Burrows et al. 2008, p. 13.
53. Housecroft & Sharpe 2008, p. 2.
54. Burrows et al. 2009, p. 110.
55. Burrows et al. 2008, p. 12.
56. "IUPAC Nomenclature of Organic Chemistry". Acdlabs.com. Retrieved 2011-06-12.
57. IUPAC Provisional Recommendations for the Nomenclature of Inorganic Chemistry (2004) .[dead link]
58. Hill, J.W.; Petrucci, R.H.; McCreary, T.W.; Perry, S.S. (2005). General Chemistry (4th ed.). Upper Saddle River, New Jersey: Pearson Prentice Hall. p. 37.
59. M. M. Avedesian; Hugh Baker. Magnesium and Magnesium Alloys. ASM International. p. 59.
60. "Official SI Unit definitions". Bipm.org. Retrieved 2011-06-12.
61. Burrows et al. 2008, p. 16.
62. Atkins & de Paula 2009, p. 9.
63. Visionlearning. "Chemical Bonding by Anthony Carpi, Ph". visionlearning. Retrieved 2011-06-12.
64. Reilly, Michael. (2007). Mechanical force induces chemical reaction, NewScientist.com news service, Reilly
65. Changing States of Matter - Chemforkids.com
66. Chemical Reaction Equation- IUPAC Goldbook
67. Gold Book Chemical Reaction IUPAC Goldbook
68. "The Lewis Acid-Base Concept". Apsidium. May 19, 2003. Archived from the original on 2008-05-27. Retrieved 2010-07-31

69. "History of Acidity". Bbc.co.uk. 2004-05-27. Retrieved 2011-06-12.
70. W.G. Laidlaw; D.E. Ryan And Gary Horlick; H.C. Clark, Josef Takats, And Martin Cowie; R.U. Lemieux (1986-12-10). "Chemistry Subdisciplines". The Canadian Encyclopedia. Retrieved 2011-06-12.
71. Herbst, Eric (May 12, 2005). "Chemistry of Star-Forming Regions". Journal of Physical Chemistry A. 109 (18): 4017–4029. doi:10.1021/jp050461c. PMID 16833724.
72. Tullo, Alexander H. (28 July 2014). "C&EN›s Global Top 50 Chemical Firms For 2014". Chemical & Engineering News. American Chemical Society. Retrieved 22 August 2014.

CHAPTER 2

Atomic Theory

Chapter Contents

- **Introduction**
- **Atomism**
- **Kinetic Theory Of Gases**
- **Physicalism**

Atomic theory is the idea that matter is made up of little units called atoms. When the ancient Greek philosopher Democritus came up with the idea in the 5th century BC, is was originally meant to refer to indivisible units. As of 1897, the British scientist J.J. Thomson discovered that atoms are in fact made up of smaller particles. Today, this theory refers to matter being made up of units that are indivisible only some of the time. Exceptions include plasmas, such as fire, other ionic arrangements, such as those found in the body, radioactive materials, and many more.

INTRODUCTION

"Atomic model" redirects here. For the unrelated term in mathematical logic, see Atomic model (mathematical logic).

This article is about the historical models of the atom. For a history of the study of how atoms combine to form molecules, see History of molecular theory.

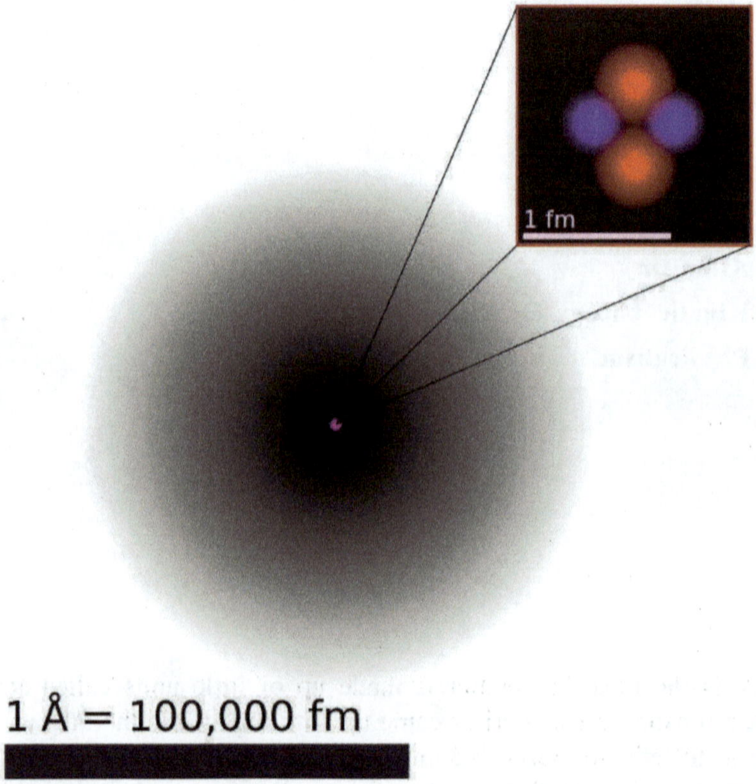

The current theoretical model of the atom involves a dense nucleus surrounded by a probabilistic "cloud" of electrons

In chemistry and physics, atomic theory is a scientific theory of the nature of matter, which states that matter is composed of discrete units called atoms. It began as a philosophical concept in ancient Greece and entered the scientific mainstream in the early 19th century when discoveries in the field of chemistry showed that matter did indeed behave as if it were made up of atoms.

The word *atom* comes from the Ancient Greek adjective *atomos*, meaning "indivisible". 19th century chemists began using the term in connection with the growing number of irreducible chemical elements. While seemingly apropos, around the turn of the 20th century, through various experiments with electromagnetism and radioactivity, physicists discovered that the so-called "uncuttable atom" was actually a conglomerate of various subatomic particles (chiefly, electrons, protons and neutrons) which can exist separately from each other. In fact, in certain extreme environments, such as neutron stars, extreme temperature and pressure prevents atoms from existing at all. Since atoms were found to be divisible, physicists later invented the term "elementary particles" to describe the "uncuttable", though not indestructible, parts of an atom. The field of science which studies subatomic particles is particle physics, and it is in this field that physicists hope to discover the true fundamental nature of matter.

History

Philosophical Atomism

The idea that matter is made up of discrete units is a very old one, appearing in many ancient cultures such as Greece and India. However, these ideas were founded in philosophical and theological reasoning rather than evidence and experimentation. Because of this, they could not convince everybody, so atomism was but one of a number of competing theories on the nature of matter. It was not until the 19th century that the idea was embraced and refined by scientists, as the blossoming science of chemistry produced discoveries that could easily be explained using the concept of atoms.

John Dalton

Near the end of the 18th century, two laws about chemical reactions emerged without referring to the notion of an atomic theory. The first was the law of conservation of mass, formulated by Antoine Lavoisier in 1789, which states that the total mass in a chemical reaction remains constant (that is, the reactants have the same mass as the products). The second was the law of definite proportions. First proven by the French chemist Joseph Louis Proust in 1799, this law states that if a compound is broken down into its constituent elements, then the masses of the constituents will always have the same proportions, regardless of the quantity or source of the original substance.

John Dalton studied and expanded upon this previous work and developed the law of multiple proportions: if two elements can be combined to form a number of possible compounds, then the ratios of the masses of the second element which combine with a fixed mass of the first element will be ratios of small whole numbers. For example: Proust had studied tin oxides and found that their masses were either 88.1% tin and 11.9% oxygen or 78.7% tin and 21.3% oxygen (these were tin(II) oxide and tin dioxide respectively). Dalton noted from these percentages that 100g of tin will combine either with 13.5g or 27g of oxygen; 13.5 and 27 form a ratio of 1:2. Dalton found that an atomic theory of matter could elegantly explain this common pattern in chemistry. In the case of Proust›s tin oxides, one tin atom will combine with either one or two oxygen atoms.

Dalton believed atomic theory could explain why water absorbed different gases in different proportions - for example, he found that water absorbed carbon dioxide far better than it absorbed nitrogen. Dalton hypothesized this was due to the differences in mass and complexity of the gases' respective particles. Indeed, carbon dioxide molecules (CO_2) are heavier and larger than nitrogen molecules (N_2).

Dalton proposed that each chemical element is composed of atoms of a single, unique type, and though they cannot be altered or destroyed by chemical means, they can combine to form more complex structures (chemical compounds). This marked the first truly scientific theory of the atom, since Dalton reached his conclusions by experimentation and examination of the results in an empirical fashion.

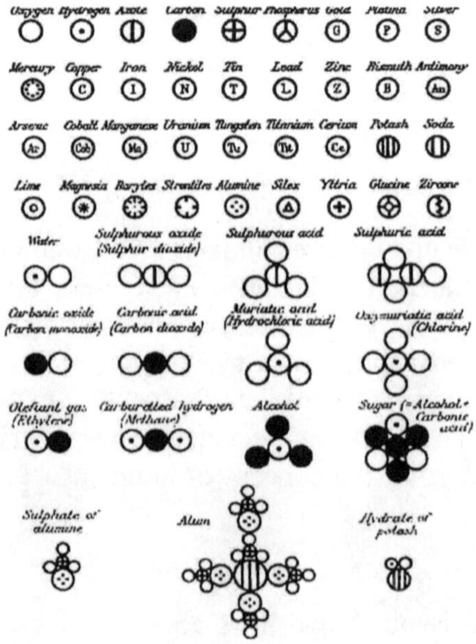

Various atoms and molecules as depicted in John Dalton's *A New System of Chemical Philosophy* (1808).

In 1803 Dalton orally presented his first list of relative atomic weights for a number of substances. This paper was published in 1805, but he did not discuss there exactly how he obtained these figures. The method was first revealed in 1807 by his acquaintance Thomas Thomson, in the third edition of Thomson›s textbook, *A System of Chemistry*. Finally, Dalton published a full account in his own textbook, *A New System of Chemical Philosophy*, 1808 and 1810.

Dalton estimated the atomic weights according to the mass ratios in which they combined, with the hydrogen atom taken as unity. However, Dalton did not conceive that with some elements atoms exist in molecules—e.g. pure oxygen exists as O2. He also mistakenly believed that the simplest compound between any two elements is always one atom of each (so he thought water was HO, not H2O). This, in addition to the crudity of his equipment, flawed his results. For instance, in 1803 he believed that oxygen atoms were 5.5 times heavier than hydrogen atoms, because in water he measured 5.5 grams of oxygen for every 1 gram of hydrogen and believed the formula for water was HO. Adopting better data, in 1806 he concluded that the atomic weight of oxygen must actually be 7 rather than 5.5, and he retained this weight for the rest of his life. Others at this time had already concluded that the oxygen atom must weigh 8 relative to hydrogen equals 1, if one assumes Dalton's formula for the water molecule (HO), or 16 if one assumes the modern water formula (H2O).

Avogadro

The flaw in Dalton's theory was corrected in principle in 1811 by Amedeo Avogadro. Avogadro had proposed that equal volumes of any two gases, at equal temperature and pressure, contain equal numbers of molecules (in other words, the mass of a gas's particles does not affect the volume that it occupies). Avogadro›s law allowed him to deduce the diatomic nature of numerous gases by studying the volumes at which they reacted. For instance: since two liters of hydrogen will react with just one liter of oxygen to produce two liters of water vapor (at constant pressure and temperature), it meant a single oxygen molecule splits in two in order to form two particles of water. Thus, Avogadro was able to offer more accurate estimates of the atomic mass of oxygen and various other elements, and made a clear distinction between molecules and atoms.

Brownian Motion

In 1827, the British botanist Robert Brown observed that dust particles inside pollen grains floating in water constantly jiggled about for no apparent reason. In 1905, Albert Einstein theorized that this Brownian motion was caused by the water molecules continuously knocking the grains about, and developed a hypothetical mathematical model to describe it. This model was validated experimentally in 1908 by French physicist Jean Perrin, thus providing additional validation for particle theory (and by extension atomic theory).

Discovery of Subatomic Particles

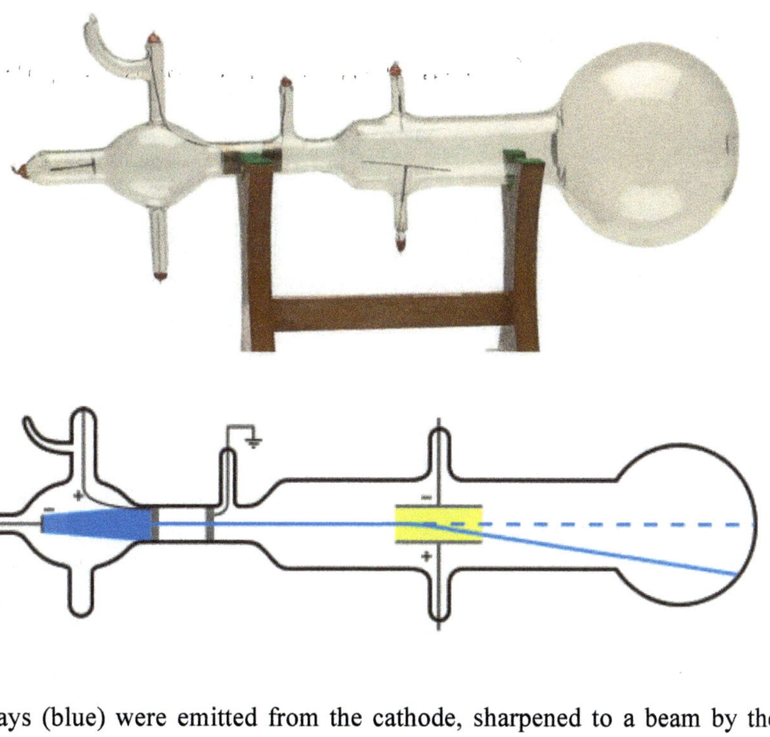

The cathode rays (blue) were emitted from the cathode, sharpened to a beam by the slits, then deflected as they passed between the two electrified plates.

Atoms were thought to be the smallest possible division of matter until 1897 when J.J. Thomson discovered the electron through his work on cathode rays.

A Crookes tube is a sealed glass container in which two electrodes are separated by a vacuum. When a voltage is applied across the electrodes, cathode rays are generated, creating a glowing patch where they strike the glass at the opposite end of the tube. Through experimentation, Thomson discovered that the rays could be deflected by an electric field (in addition to magnetic fields, which was already known). He concluded that these rays, rather than being a form of light, were composed of very light negatively charged particles he called "corpuscles" (they would later be renamed electrons by other scientists). He measured the mass-to-charge ratio and discovered it was 1800 times smaller than that of hydrogen, the smallest atom. These corpuscles were a particle unlike any other previously known.

Thomson suggested that atoms were divisible, and that the corpuscles were their building blocks. To explain the overall neutral charge of the atom, he proposed that the corpuscles were distributed in a uniform sea of positive charge; this was the plum pudding model as the electrons were embedded in the positive charge like plums in a plum pudding (although in Thomson›s model they were not stationary).

Discovery of the Nucleus

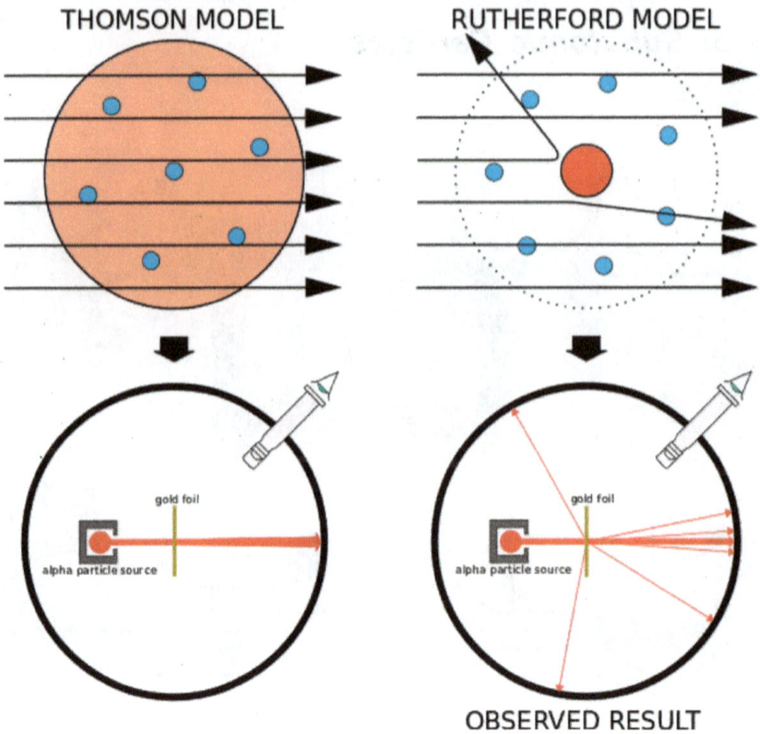

The Geiger-Marsden Experiment

Left: Expected results: alpha particles passing through the plum pudding model of the atom with negligible deflection.

Right: Observed results: a small portion of the particles were deflected by the concentrated positive charge of the nucleus.

Thomson›s plum pudding model was disproved in 1909 by one of his former students, Ernest Rutherford, who discovered that most of the mass and positive charge of an atom is concentrated in a very small fraction of its volume, which he assumed to be at the very center.

In the Geiger–Marsden experiment, Hans Geiger and Ernest Marsden (colleagues of Rutherford working at his behest) shot alpha particles at thin sheets of metal and measured their deflection through the use of a fluorescent screen. Given the very small mass of the electrons, the high momentum of the alpha particles, and the low concentration of the positive charge of the plum pudding model, the experimenters expected all the alpha particles to pass through the metal foil without significant deflection. To their astonishment, a small fraction of the alpha particles experienced heavy deflection. Rutherford concluded that the positive charge of the atom must be concentrated in a very tiny volume to produce an electric field sufficiently intense to deflect the alpha particles so strongly.

This led Rutherford to propose a planetary model in which a cloud of electrons surrounded a small, compact nucleus of positive charge. Only such a concentration of charge could produce the electric field strong enough to cause the heavy deflection.

First Steps toward a Quantum Physical Model of the Atom

The planetary model of the atom had two significant shortcomings. The first is that, unlike planets orbiting a sun, electrons are charged particles. An accelerating electric charge is known to emit electromagnetic waves according to the Larmor formula in classical electromagnetism. An orbiting charge should steadily lose energy and spiral toward the nucleus, colliding with it in a small fraction of a second. The second problem was that the planetary model could not explain the highly peaked emission and absorption spectra of atoms that were observed.

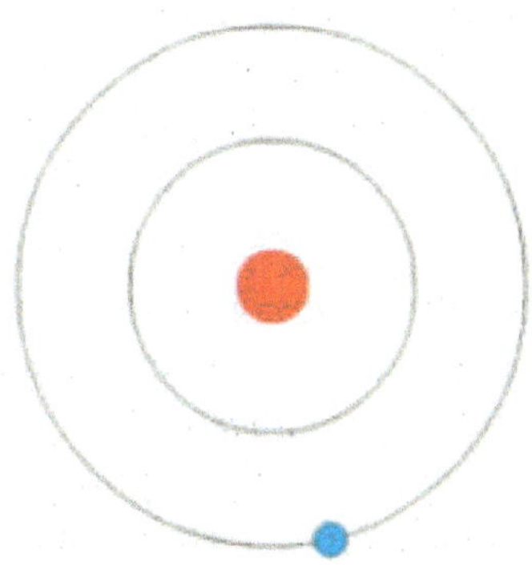

The Bohr model of the atom

Quantum theory revolutionized physics at the beginning of the 20th century, when Max Planck and Albert Einstein postulated that light energy is emitted or absorbed in discrete amounts known as quanta (singular, *quantum*). In 1913, Niels Bohr incorporated this idea into his Bohr model of the atom, in which an electron could only orbit the nucleus in particular circular orbits with fixed angular momentum and energy, its distance from the nucleus (i.e., their radii) being proportional to its energy. Under this model an electron could not spiral into the nucleus because it could not lose energy in a continuous manner; instead, it could only make instantaneous «quantum leaps» between the fixed energy levels. When this occurred, light was emitted or absorbed at a frequency proportional to the change in energy (hence the absorption and emission of light in discrete spectra).

Bohr›s model was not perfect. It could only predict the spectral lines of hydrogen; it couldn't predict those of multielectron atoms. Worse still, as spectrographic technology improved, additional spectral lines in hydrogen were observed which Bohr's model couldn't explain. In 1916, Arnold Sommerfeld added elliptical orbits to the Bohr model to explain the extra emission lines, but this made the model very difficult to use, and it still couldn›t explain more complex atoms.

Discovery of Isotopes

While experimenting with the products of radioactive decay, in 1913 radiochemist Frederick Soddy discovered that there appeared to be more than one element at each position on the periodic table. The term isotope was coined by Margaret Todd as a suitable name for these elements.

That same year, J.J. Thomson conducted an experiment in which he channeled a stream of neon ions through magnetic and electric fields, striking a photographic plate at the other end. He observed two glowing patches on the plate, which suggested two different deflection trajectories. Thomson concluded this was because some of the neon ions had a different mass. The nature of this differing mass would later be explained by the discovery of neutrons in 1932.

Discovery of Nuclear Particles

In 1917 Rutherford bombarded nitrogen gas with alpha particles and observed hydrogen nuclei being emitted from the gas (Rutherford recognized these, because he had previously obtained them bombarding hydrogen with alpha particles, and observing hydrogen nuclei in the products). Rutherford concluded that the hydrogen nuclei emerged from the nuclei of the nitrogen atoms themselves (in effect, he had split a nitrogen).

From his own work and the work of his students Bohr and Henry Moseley, Rutherford knew that the positive charge of any atom could always be equated to that of an integer number of hydrogen nuclei. This, coupled with the atomic mass of many elements being roughly equivalent to an integer number of hydrogen atoms - then assumed to be the lightest particles - led him to conclude that hydrogen nuclei were singular particles and a basic constituent of all atomic nuclei. He named such particles protons. Further experimentation by Rutherford found that the nuclear mass of most atoms exceeded that of the protons it possessed; he speculated that this surplus mass was composed of hitherto unknown neutrally charged particles, which were tentatively dubbed "neutrons".

In 1928, Walter Bothe observed that beryllium emitted a highly penetrating, electrically neutral radiation when bombarded with alpha particles. It was later discovered that this radiation could knock hydrogen atoms out of paraffin wax. Initially it was thought to be high-energy gamma radiation, since gamma radiation had a similar effect on electrons in metals, but James Chadwick found that the ionization effect was too strong for it to be due to electromagnetic radiation, so long as energy and momentum were conserved in the interaction. In 1932, Chadwick exposed various elements, such as hydrogen and nitrogen, to the mysterious «beryllium radiation», and by measuring the energies of the recoiling charged particles, he deduced that the radiation was actually composed of electrically neutral particles which could not be massless like the gamma ray, but instead were required to have a mass similar to that of a proton. Chadwick now claimed these particles as Rutherford›s neutrons. For his discovery of the neutron, Chadwick received the Nobel Prize in 1935.

Quantum Physical Models of the Atom

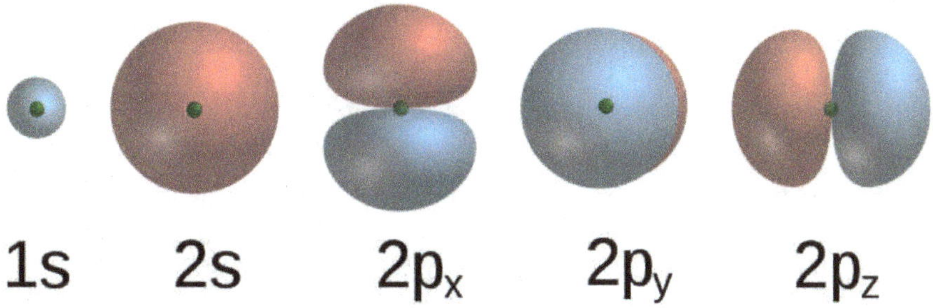

The five filled atomic orbitals of a neon atom separated and arranged in order of increasing energy from left to right, with the last three orbitals being equal in energy. Each orbital holds up to two electrons, which most probably exist in the zones represented by the colored bubbles. Each electron is equally present in both orbital zones, shown here by color only to highlight the different wave phase.

In 1924, Louis de Broglie proposed that all moving particles—particularly subatomic particles such as electrons—exhibit a degree of wave-like behavior. Erwin Schrödinger, fascinated by this idea, explored whether or not the movement of an electron in an atom could be better explained as a wave rather than as a particle. Schrödinger's equation, published in 1926, describes an electron as a wavefunction instead of as a point particle. This approach elegantly predicted many of the spectral phenomena that Bohr›s model failed to explain. Although this concept was mathematically convenient, it was difficult to visualize, and faced opposition. One of its critics, Max Born, proposed instead that Schrödinger›s wavefunction described not the electron but rather all its possible states, and thus could be used to calculate the probability of finding an electron at any given location around the nucleus. This reconciled the two opposing theories of particle versus wave electrons and the idea of wave–particle duality was introduced. This theory stated that the electron may exhibit the properties of both a wave and a particle. For example, it can be refracted like a wave, and has mass like a particle.

A consequence of describing electrons as waveforms is that it is mathematically impossible to simultaneously derive the position and momentum of an electron. This became known as the Heisenberg uncertainty principle after the theoretical physicist Werner Heisenberg, who first described it and published it in 1927. This invalidated Bohr's model, with its neat, clearly defined circular orbits. The modern model of the atom describes the positions of electrons in an atom in terms of probabilities. An electron can potentially be found at any distance from the nucleus, but, depending on its energy level, exists more frequently in certain regions around the nucleus than others; this pattern is referred to as its atomic orbital. The orbitals come in a variety of shapes-sphere, dumbbell, torus, etc.-with the nucleus in the middle.

ATOMISM

Atomism (from Greek τομον, atomon, i.e. "uncuttable", "indivisible") is a natural philosophy that developed in several ancient traditions. The atomists theorized that nature consists of two fundamental principles: atom and void. Unlike their modern scientific namesake in atomic theory, philosophical atoms come in an infinite variety of shapes and sizes, each indestructible, immutable and surrounded by a void where they collide with the others or hook together forming a cluster. Clusters of different shapes, arrangements, and positions give rise to the various macroscopic substances in the world.

References to the concept of atomism and its atoms are found in ancient India and ancient Greece. In the West, atomism emerged in the 5th century BCE with Leucippus and Democritus. In India the Jain, Ajivika and Carvaka schools of atomism may date back to the 4th century BCE. The Nyaya and Vaisheshika schools later developed theories on how atoms combined into more complex objects. Whether Indian culture influenced Greek or vice versa or whether both evolved independently is a matter of dispute.

The particles of chemical matter for which chemists and other natural philosophers of the early 19th century found experimental evidence were thought to be indivisible, and therefore were given the name "atom", long used by the atomist philosophy.

However, in the 20th century, the "atoms" of the chemists were found to be composed of even smaller entities: electrons, neutrons and protons, and further experiments showed that protons and neutrons are made of quarks. Although the connection to historical atomism is at best tenuous, elementary particles have thus become a modern analog of philosophical atoms, despite the misnomer in chemistry.

Reductionism

Philosophical atomism is a reductive argument: not only that everything is composed of atoms and void, but that nothing they compose really exists: the only things that really exist are atoms ricocheting off each other mechanistically in an otherwise empty void. Atomism stands in contrast to a substance theory wherein a prime material continuum remains qualitatively invariant under division (for example, the ratio of the four classical elements would be the same in any portion of a homogeneous material).

Indian Buddhists, such as Dharmakirti and others, also developed distinctive theories of atomism, for example, involving momentary (instantaneous) atoms, that flash in and out of existence (Kalapas).

Greek Atomism

In the 5th century B.C.E., Leucippus and his pupil Democritus proposed that all matter was composed of small indivisible particles called atoms, in order to reconcile two conflicting schools of thought on the nature of reality. On one side was Heraclitus, who believed that the nature of all existence is change. On the other side was Parmenides, who believed instead that all change is illusion.

Parmenides denied the existence of motion, change and void. He believed all existence to be a single, all-encompassing and unchanging mass (a concept known as monism), and that change and motion were mere illusions. This conclusion, as well as the reasoning that led to it, may indeed seem baffling to the modern empirical mind, but Parmenides explicitly rejected sensory experience as the path to an understanding of the universe, and instead used purely abstract reasoning. Firstly, he believed there is no such thing as void, equating it with non-being (i.e. "if the void is, then it is not nothing; therefore it is not the void"). This in turn meant that motion is impossible, because there is no void to move into. He also wrote all that is must be an indivisible unity, for if it were manifold, then there would have to be a void that could divide it (and he did not believe the void exists). Finally, he stated that the all encompassing Unity is unchanging, for the Unity already encompasses all that is and can be.

Democritus accepted most of Parmenides' arguments, except for the idea that change is an illusion. He believed change was real, and if it was not then at least the illusion had to be explained. He thus supported the concept of void, and stated that the universe is made up of many Parmenidean entities that move around in the void. The void is infinite and provides the space in which the atoms can pack or scatter differently. The different possible packings and scatterings within the void make up the shifting outlines and bulk of the objects that organisms feel, see, eat, hear, smell, and taste. While organisms may feel hot or cold, hot and cold actually have no real existence. They are simply sensations produced in organisms by the different packings and scatterings of the atoms in the void that compose the object that organisms sense as being "hot" or "cold".

The work of Democritus only survives in secondhand reports, some of which are unreliable or conflicting. Much of the best evidence of Democritus' theory of atomism is reported by Aristotle in his discussions of Democritus' and Plato's contrasting views on the types of indivisibles composing the natural world.

Geometry and Atoms

Plato (c. 427 — c. 347 B.C.E.), were he familiar with the atomism of Democritus, would have objected to its mechanistic materialism. He argued that atoms just crashing into other atoms could never produce the beauty and form of the world. In Plato's Timaeus, (28B – 29A) the character of Timeaus insisted that the cosmos was not eternal but was created, although its creator framed it after an eternal, unchanging model.

One part of that creation were the four simple bodies of fire, air, water, and earth. But Plato did not consider these corpuscles to be the most basic level of reality, for in his view they were made up of an unchanging level of reality, which was mathematical. These simple bodies were geometric solids, the faces of which were, in turn, made up of triangles. The square faces of the cube were each made up of four isosceles right-angled triangles and the triangular faces of the tetrahedron, octahedron, and icosahedron were each made up of six right-angled triangles.

Element	Polyhedron		Number of Faces	Number of Triangles
Fire	Tetrahedron (Animation)		4	24
Air	Octahedron (Animation)		8	48
Water	Icosahedron (Animation)		20	120
Earth	Cube (Animation)		6	24

Geometrical Simple Bodies According to Plato

He postulated the geometric structure of the simple bodies of the four elements as summarized in the adjacent table. The cube, with its flat base and stability, was assigned to earth; the tetrahedron was assigned to fire because its penetrating points and sharp edges made it mobile. The points and edges of the octahedron and icosahedron were blunter and so these less mobile bodies were assigned to air and water. Since the simple bodies could be decomposed into triangles, and the triangles reassembled into atoms of different elements, Plato's model offered a plausible account of changes among the primary substances.

The Rejection of Atoms

Sometime before 330 B.C.E. Aristotle asserted that the elements of fire, air, earth, and water were not made of atoms, but were continuous. Aristotle considered the existence of a void, which was required by atomic theories, to violate physical principles. Change took place not by the rearrangement of atoms to make new structures, but by transformation of matter from what it was in potential to a new actuality. A piece of wet clay, when acted upon by a potter, takes on its potential to be an actual drinking mug. Aristotle has often been criticized for rejecting atomism, but in ancient Greece the atomic theories of

Democritus remained "pure speculations, incapable of being put to any experimental test. Granted that atomism was, in the long run, to prove far more fruitful than any qualitative theory of matter, in the short run the theory that Aristotle proposed must have seemed in some respects more promising".

Later Ancient Atomism

Epicurus (341–270) studied atomism with Nausiphanes who had been a student of Democritus. Although Epicurus was certain of the existence of atoms and the void, he was less sure we could adequately explain specific natural phenomena such as earthquakes, lightning, comets, or the phases of the Moon (Lloyd 1973, 25–6). Few of Epicurus's writings survive and those that do reflect his interest in applying Democritus's theories to assist people in taking responsibility for themselves and for their own happiness—since he held there are no gods around that can help them. He understood gods' role as moral ideals.

His ideas are also represented in the works of his follower Lucretius, who wrote On the Nature of Things. This scientific work in poetic form illustrates several segments of Epicurean theory on how the universe came into its current stage and it shows that the phenomena we perceive are actually composite forms. The atoms and the void are eternal and in constant motion. Atomic collisions create objects, which are still composed of the same eternal atoms whose motion for a while is incorporated into the created entity. Human sensations and meteorological phenomena are also explained by Lucretius in terms of atomic motion.

Atomism and Ethics

Some later philosophers attributed the idea, that man created gods and the gods did not create man, to Democritus. For example, Sextus Empiricus noted:

Some people think that we arrived at the idea of gods from the remarkable things that happen in the world. Democritus ... says that the people of ancient times were frightened by happenings in the heavens such as thunder, lightning, ..., and thought that they were caused by gods.

Three hundred years after Epicurus, Lucretius in his epic poem On the Nature of Things would depict him as the hero who crushed the monster Religion through educating the people in what was possible in the atoms and what was not possible in the atoms. However, Epicurus expressed a non-aggressive attitude characterized by his statement: "The man who best knows how to meet external threats makes into one family all the creatures he can; and those he can not, he at any rate does not treat as aliens; and where he finds even this impossible, he avoids all dealings, and, so far as is advantageous, excludes them from his life."

Indian Atomism

The Indian atomistic position, like many movements in Indian Philosophy and Mathematics, starts with an argument from Linguistics. The Vedic etymologist and grammarian Yaska

(c. 7th century BC) in his Nirukta, in dealing with models for how linguistic structures get to have their meanings, takes the atomistic position that words are the "primary" carrier of meaning – i.e. words have a preferred ontological status in defining meaning. This position was to be the subject of a fierce debate in the Indian tradition from the early Christian era till the 18th century, involving different philosophers from the Nyaya, Mimamsa and Buddhist schools.

In the pratishakhya text (c. 2nd century BCE), the gist of the controversy was stated cryptically in the sutra form as "saMhitA pada-prakr^tiH". According to the atomist view, the words (pada) would be the primary elements (prakrti) out of which the sentence is constructed, while the holistic view considers the sentence as the primary entity, originally "given" in its context of utterance, and the words are arrived at only through analysis and abstraction.

These two positions came to be called a-kShaNDa-pakSha (indivisibility or sentence-holism), a position developed later by Bhartrihari (c. 500 AD), vs. kShaNDa-pakSha (atomism), a position adopted by the Mimamsa and Nyaya schools (Note: kShanDa = fragmented; "a-kShanDa" = whole).

Between the 5th and 3rd centuries BC, the atom (anu or aor) is mentioned in the Bhagavad Gita (Chapter 8, Verse 9):

*kaviṁ purāṇam anuśāsitāram **aṇor** aṇīyāṁsam anusmared*

yaḥ sarvasya dhātāram acintya-rūpam āditya-varṇaṁ tamasaḥ parastāt

One meditates on the omniscient, primordial, the controller, smaller than the atom, yet the maintainer of everything; whose form is inconceivable, resplendent like the sun and totally transcendental to material nature

The ancient shāshvata-vāda doctrine of eternalism, which held that elements are eternal, is also suggestive of a possible starting point for atomism (Gangopadhyaya 1981).

The atomist position had transcended language into epistemology by the time that Nyaya–Vaisesika, Buddhist and Jaina theology were developing mature philosophical positions.

Will Durant wrote in Our Oriental Heritage:

"Two systems of Indian thought propound physical theories suggestively similar to those of Greece. Kanada, founder of the Vaisheshika philosophy, held that the world was composed of atoms as many in kind as the various elements. The Jains more nearly approximated to Democritus by teaching that all atoms were of the same kind, producing different effects by diverse modes of combinations. Kanada believed light and heat to be varieties of the same substance; Udayana taught that all heat comes from the sun; and Vachaspati, like Newton, interpreted light as composed of minute particles emitted by substances and striking the eye."

Indian atomism in the Middle Ages was still mostly philosophical and/or religious in intent, though it was also scientific. Because the "infallible Vedas", the oldest Hindu texts, do not mention atoms (though they do mention elements), atomism was not orthodox in many schools of Hindu philosophy, although accommodationist interpretations or assumptions of lost text justified the use of atomism for non-orthodox schools of Hindu thought. The Buddhist and Jaina schools, however, were more willing to accept the ideas of atomism.

Nyaya–Vaisesika School

The Nyaya–Vaisesika school developed one of the earliest forms of atomism; scholars date the Nyaya and Vaisesika texts from the 6th to 1st centuries BC. Like the Buddhist atomists, the Vaisesika had a pseudo-Aristotelian theory of atomism. They posited the four elemental atom types, but in Vaisesika physics atoms had 24 different possible qualities, divided between general extensive properties and specific (intensive) properties. Like the Jaina school, the Nyaya–Vaisesika atomists had elaborate theories of how atoms combine. In both Jaina and Vaisesika atomism, atoms first combine in pairs (dyads), and then group into trios of pairs (triads), which are the smallest visible units of matter.

Buddhist School

The Buddhist atomists had very qualitative, Aristotelian-style atomic theory. According to ancient Buddhist atomism, which probably began developing before the 4th century BC, there are four kinds of atoms, corresponding to the standard elements. Each of these elements has a specific property, such as solidity or motion, and performs a specific function in mixtures, such as providing support or causing growth. Like the Hindu Jains, the Buddhists were able to integrate a theory of atomism with their theological presuppositions. Later Indian Buddhist philosophers, such as Dharmakirti and Dignāga, considered atoms to be point-sized, durationless, and made of energy.

Jaina School

The most elaborate and well-preserved Indian theory of atomism comes from the philosophy of the Jaina school, dating back to at least the 6th century BC. Some of the Jain texts that refer to matter and atoms are Pancastikayasara, Kalpasutra, Tattvarthasutra and Pannavana Suttam. The Jains envisioned the world as consisting wholly of atoms, except for souls. Paramāṇus or atoms were considered as the basic building blocks of all matter. Their concept of atoms was very similar to classical atomism, differing primarily in the specific properties of atoms. Each atom, according to Jain philosophy, has one kind of taste, one smell, one color, and two kinds of touch, though it is unclear what was meant by "kind of touch". Atoms can exist in one of two states: subtle, in which case they can fit in infinitesimally small spaces, and gross, in which case they have extension and occupy a finite space. Certain characteristics of Paramāṇu correspond with that sub-atomic particles. For example, Paramāṇu is characterized by continuous motion either in a straight line or in case of attractions from other Paramāṇus, it follows a curved path. This corresponds with the description of orbit of electrons across the Nucleus. Ultimate particles are also described as particles with positive (Snigdha i.e. smooth charge) and negative (Rūksa – rough) charges that provide them the binding force. Although atoms are made of the same basic substance, they can combine based on their eternal properties to produce any of six "aggregates", which seem to correspond with the Greek concept of "elements": earth, water, shadow, sense objects, karmic matter, and unfit matter. To the Jains, karma was real, but was a naturalistic, mechanistic phenomenon caused by buildups of subtle karmic matter within the soul. They also had detailed theories of how atoms could combine, react, vibrate, move, and perform other actions, all of which were thoroughly deterministic.

Time as Explained in Srimad-Bhagavatam

"The material manifestation's ultimate particle, which is indivisible and not formed into a body, is called the atom (parama-auh). It exists always as an invisible identity, even after the dissolution of all forms. The material body is but a combination of such atoms, but it is misunderstood by the common man.

Atoms are the ultimate state of the manifest universe. When they stay in their own forms without forming different bodies, they are called the unlimited oneness. There are certainly different bodies in physical forms, but the atoms themselves form the complete manifestation.

One can estimate time by measuring the movement of the atomic combination of bodies. Time is the potency of the almighty Personality of Godhead, Hari, who controls all physical movement although He is not visible in the physical world.

The division of gross time is calculated as follows: two atoms make one double atom, and three double atoms make one hexatom. This hexatom is visible in the sunshine which enters through the holes of a window screen. One can clearly see that the hexatom goes up towards the sky.

The atom is described as an invisible particle, but when six such atoms combine together, they are called a trasareu, and this is visible in the sunshine pouring through the holes of a window screen."

— Excerpted from Srimad-Bhagavatam (SB 3.11.1–5) by A. C. Bhaktivedanta Swami Prabhupada, courtesy of the Bhaktivedanta Book Trust International (prabhupadabooks.com)

Islamic Atomism

Atomistic philosophies are found very early in Islamic philosophy and was influenced by earlier Greek and to some extent Indian philosophy. Like both the Greek and Indian versions, Islamic atomism was a charged topic that had the potential for conflict with the prevalent religious orthodoxy, but it was instead more often favoured by orthodox Islamic theologians. It was such a fertile and flexible idea that, as in Greece and India, it flourished in some leading schools of Islamic thought.

Asharite Atomism

The most successful form of Islamic atomism was in the Asharite school of Islamic theology, most notably in the work of the theologian al-Ghazali (1058–1111). In Asharite atomism, atoms are the only perpetual, material things in existence, and all else in the world is "accidental" meaning something that lasts for only an instant. Nothing accidental can be the cause of anything else, except perception, as it exists for a moment. Contingent events are not subject to natural physical causes, but are the direct result of God's constant intervention, without which nothing could happen. Thus nature is completely dependent on God, which meshes with other Asharite Islamic ideas on causation, or the lack thereof (Gardet 2001). Al-Ghazali also used the theory to support his theory of occasionalism. In a sense, the Asharite theory of atomism has far more in common with Indian atomism than it does with Greek atomism.

Averroism

Other traditions in Islam rejected the atomism of the Asharites and expounded on many Greek texts, especially those of Aristotle. An active school of philosophers in Spain, including the noted commentator Averroes (AD 1126–1198) explicitly rejected the thought of al-Ghazali and turned to an extensive evaluation of the thought of Aristotle. Averroes commented in detail on most of the works of Aristotle and his commentaries did much to guide the interpretation of Aristotle in later Jewish and Christian scholastic thought.

Medieval European Speculations

While Aristotelian philosophy eclipsed the importance of the atomists in late Roman and medieval Europe, their work was still preserved and exposited through commentaries on the works of Aristotle. In the 2nd century, Galen (AD 129–216) presented extensive discussions of the Greek atomists, especially Epicurus, in his Aristotle commentaries. According to historian of atomism Joshua Gregory, there was no serious work done with atomism from the time of Galen until Gassendi and Descartes resurrected it in the 17th century; "the gap between these two 'modern naturalists' and the ancient Atomists marked "the exile of the atom" and "it is universally admitted that the Middle Ages had abandoned Atomism, and virtually lost it."

However, although the ancient atomists' works were unavailable, Scholastic thinkers still had Aristotle's critiques of atomism. In the medieval universities there were expressions of atomism. For example, in the 14th century Nicholas of Autrecourt considered that matter, space, and time were all made up of indivisible atoms, points, and instants and that all generation and corruption took place by the rearrangement of material atoms. The similarities of his ideas with those of al-Ghazali suggest that Nicholas may have been familiar with Ghazali's work, perhaps through Averroes' refutation of it (Marmara, 1973–74).

Scholastic Minima Naturalia

Although the atomism of Epicurus had fallen out of favor in the centuries of Scholasticism, a related Aristotelian concept, that of minima naturalia (natural minima) received extensive consideration. Minima naturalia were theorized by Aristotle as the smallest parts into which a homogeneous natural substance (e.g., flesh, bone, or wood) could be divided and still retain its essential character. Speculation on minima naturalia provided philosophical background for the mechanistic philosophy of early modern thinkers like Descartes, and for the alchemical works of Geber and Daniel Sennert, who in turn influenced the corpuscularian alchemist Robert Boyle, one of the founders of modern chemistry.

Unlike the atomism of Leucippus, Democritus, and Epicurus, and also unlike the later atomic theory of John Dalton, the Aristotelian natural minimum was not conceptualized as physically indivisible--"atomic" in the contemporary sense. Instead, the concept was rooted in Aristotle's hylomorphic worldview, which held that every physical thing is a compound of matter (Greek hyle) and an immaterial substantial form (Greek morphe) that imparts its essential nature and structure. For instance, a rubber ball for a hylomorphist like Aristotle would be rubber (matter) structured by spherical shape (form).

Aristotle's intuition was that there is some smallest size beyond which matter could no longer be structured as flesh, or bone, or wood, or some other such organic substance that for Aristotle, living before the microscope, could be considered homogeneous. For instance, if flesh were divided beyond its natural minimum, what would be left might be a large amount of the element water, and smaller amounts of the other elements. But whatever water or other elements were left, they would no longer have the "nature" of flesh: in hylomorphic terms, they would no longer be matter structured by the form of flesh; instead the remaining water, e.g., would be matter structured by the form of water, not the form of flesh. This is suggestive of modern chemistry, in which, e.g., a bar of gold can be continually divided until one has a single atom of gold, but further division yields only subatomic particles (electrons, quarks, etc.) which are no longer "gold." However, the parallel is not exact: minima naturalia are not a direct anticipation of modern chemical and physical concepts.

A chief theme in late Roman and Scholastic commentary on this concept is reconciling minima naturalia with the general Aristotelian principle of infinite divisibility. Commentators like John Philoponus and Thomas Aquinas reconciled these aspects of Aristotle's thought by distinguishing between mathematical and "natural" divisibility. With few exceptions, much of the curriculum in the universities of Europe was based on such Aristotelianism for most of the Middle Ages (Kargon 1966). Scholasticism was standard science in the time of Isaac Newton, but in the 17th century, a renewed interest in Epicurean atomism and corpuscularianism as a hybrid or an alternative to Aristotelian physics had begun to mount outside the classroom.

Atomic Renaissance

The main figures in the rebirth of atomism were René Descartes, Pierre Gassendi, and Robert Boyle, as well as other notable figures.

One of the first groups of atomists in England was a cadre of amateur scientists known as the Northumberland circle, led by Henry Percy (1585–1632), the 9th Earl of Northumberland. Although they published little of account, they helped to disseminate atomistic ideas among the burgeoning scientific culture of England, and may have been particularly influential to Francis Bacon, who became an atomist around 1605, though he later rejected some of the claims of atomism. Though they revived the classical form of atomism, this group was among the scientific avant-garde: the Northumberland circle contained nearly half of the confirmed Copernicans prior to 1610 (the year of Galileo's The Starry Messenger). Other influential atomists of late 16th and early 17th centuries include Giordano Bruno, Thomas Hobbes (who also changed his stance on atomism late in his career), and Thomas Hariot. A number of different atomistic theories were blossoming in France at this time, as well (Clericuzio 2000).

Galileo Galilei (1564–1642) was an advocate of atomism in his 1612, Discourse on Floating Bodies (Redondi 1969). In The Assayer, Galileo offered a more complete physical system based on a corpuscular theory of matter, in which all phenomena—with the exception of sound—are produced by "matter in motion". Galileo identified some basic problems with Aristotelian physics through his experiments. He utilized a theory of atomism as a partial replacement, but he was never unequivocally committed to it. For example, his experiments with falling bodies and inclined planes led him to the concepts of circular inertial motion and accelerating free-fall. The current Aristotelian theories of impetus and terrestrial motion were inadequate to explain these. While atomism did not explain the law of fall either, it was a more promising framework in which to develop an explanation because motion was conserved in ancient atomism (unlike Aristotelian physics).

René Descartes' (1596–1650) "mechanical" philosophy of corpuscularism had much in common with atomism, and is considered, in some senses, to be a different version of it. Descartes thought everything physical in the universe to be made of tiny vortices of matter. Like the ancient atomists, Descartes claimed that sensations, such as taste or temperature, are caused by the shape and size of tiny pieces of matter. The main difference between atomism and Descartes' concept was the existence of the void. For him, there could be no vacuum, and all matter was constantly swirling to prevent a void as corpuscles moved through other matter. Another key distinction between Descartes' view and classical atomism is the mind/body duality of Descartes, which allowed for an independent realm of existence for thought, soul, and most importantly, God. Gassendi's concept was closer to classical atomism, but with no atheistic overtone.

Pierre Gassendi (1592–1655) was a Catholic priest from France who was also an avid natural philosopher. He was particularly intrigued by the Greek atomists, so he set out to "purify" atomism from its heretical and atheistic philosophical conclusions (Dijksterhius 1969). Gassendi formulated his atomistic conception of mechanical philosophy partly in response to Descartes; he particularly opposed Descartes' reductionist view that only purely mechanical explanations of physics are valid, as well as the application of geometry to the whole of physics (Clericuzio 2000).

Corpuscularianism

Corpuscularianism is similar to atomism, except that where atoms were supposed to be indivisible, corpuscles could in principle be divided. In this manner, for example, it was theorized that mercury could penetrate into metals and modify their inner structure, a step on the way towards transmutative production of gold. Corpuscularianism was associated by its leading proponents with the idea that some of the properties that objects appear to have are artifacts of the perceiving mind: 'secondary' qualities as distinguished from 'primary' qualities. Not all corpuscularianism made use of the primary-secondary quality distinction, however. An influential tradition in medieval and early modern alchemy argued that chemical analysis revealed the existence of robust corpuscles that retained their identity in chemical compounds (to use the modern term). William R. Newman has dubbed this approach to matter theory "chymical atomism," and has argued for its significance to both the mechanical philosophy and to the chemical atomism that emerged in the early

19th century. Corpuscularianism stayed a dominant theory over the next several hundred years and retained its links with alchemy in the work of scientists such as Robert Boyle and Isaac Newton in the 17th century. It was used by Newton, for instance, in his development of the corpuscular theory of light. The form that came to be accepted by most English scientists after Robert Boyle (1627–1692) was an amalgam of the systems of Descartes and Gassendi. In The Sceptical Chymist (1661), Boyle demonstrates problems that arise from chemistry, and offers up atomism as a possible explanation. The unifying principle that would eventually lead to the acceptance of a hybrid corpuscular–atomism was mechanical philosophy, which became widely accepted by physical sciences.

Atomic Theory

By the late 18th century, the useful practices of engineering and technology began to influence philosophical explanations for the composition of matter. Those who speculated on the ultimate nature of matter began to verify their "thought experiments" with some repeatable demonstrations, when they could.

Roger Boscovich provided the first general mathematical theory of atomism, based on the ideas of Newton and Leibniz but transforming them so as to provide a programme for atomic physics.

In 1808, John Dalton assimilated the known experimental work of many people to summarize the empirical evidence on the composition of matter. He noticed that distilled water everywhere analyzed to the same elements, hydrogen and oxygen. Similarly, other purified substances decomposed to the same elements in the same proportions by weight.

Therefore we may conclude that the ultimate particles of all homogeneous bodies are perfectly alike in weight, figure, etc. In other words, every particle of water is like every other particle of water; every particle of hydrogen is like every other particle of hydrogen, etc.

Furthermore, he concluded that there was a unique atom for each element, using Lavoisier's definition of an element as a substance that could not be analyzed into something simpler. Thus, Dalton concluded the following.

Chemical analysis and synthesis go no farther than to the separation of particles one from another, and to their reunion. No new creation or destruction of matter is within the reach of chemical agency. We might as well attempt to introduce a new planet into the solar system, or to annihilate one already in existence, as to create or destroy a particle of hydrogen. All the changes we can produce, consist in separating particles that are in a state of cohesion or combination, and joining those that were previously at a distance.

And then he proceeded to give a list of relative weights in the compositions of several common compounds, summarizing:

1st. That water is a binary compound of hydrogen and oxygen, and the relative weights of the two elementary atoms are as 1:7, nearly;

2nd. That ammonia is a binary compound of hydrogen and azote nitrogen, and the relative weights of the two atoms are as 1:5, nearly...

Dalton concluded that the fixed proportions of elements by weight suggested that the atoms of one element combined with only a limited number of atoms of the other elements to form the substances that he listed.

Atomic Theory Controversy

Dalton's atomic theory remained controversial throughout the 19th century. Whilst the Law of definite proportion was accepted, the hypothesis that this was due to atoms was not so widely accepted. For example, in 1826 when Sir Humphry Davy presented Dalton the Royal Medal from the Royal Society, Davy said that the theory only became useful when the atomic conjecture was ignored. Sir Benjamin Collins Brodie in 1866 published the first part of his Calculus of Chemical Operations as a non-atomic alternative to the Atomic Theory. He described atomic theory as a 'Thoroughly materialistic bit of joiners work'. Alexander Williamson used his Presidential Address to the London Chemical Society in 1869 to defend the Atomic Theory against its critics and doubters. This in turn led to further meetings at which the positivists again attacked the supposition that there were atoms. The matter was finally resolved in Dalton's favour in the early 20th century with the rise of atomic physics.

KINETIC THEORY OF GASES

The kinetic theory describe a gas as a large number of submicroscopic particles (atoms or molecules), all of which are in constant rapid motion that has randomness arising from their many collisions with each other and with the walls of the container.

Kinetic theory explains macroscopic properties of gases, such as pressure, temperature, viscosity, thermal conductivity, and volume, by considering their molecular composition and motion. The theory posits that gas pressure is due to the impacts, on the walls of a container, of molecules or atoms moving at different velocities.

Kinetic theory defines temperature in its own way, not identical with the thermodynamic definition.

Under a microscope, the molecules making up a liquid are too small to be visible, but the jittery motion of pollen grains or dust particles can be seen. Known as Brownian motion, it results directly from collisions between the grains or particles and liquid molecules. As analyzed by Albert Einstein in 1905, this experimental evidence for kinetic theory is generally seen as having confirmed the concrete material existence of atoms and molecules.

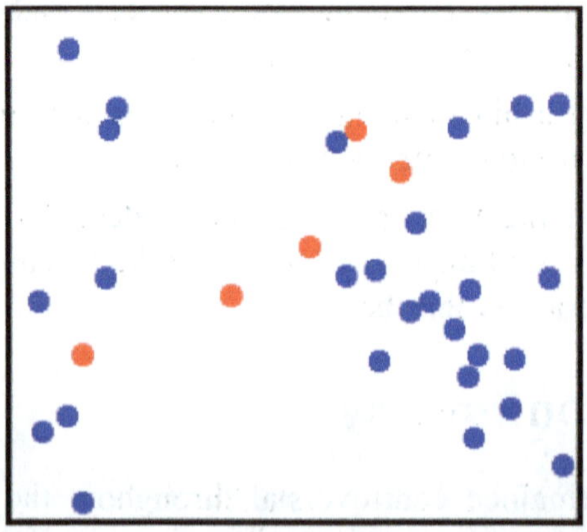

The temperature of an ideal monatomic gas is proportional to the average kinetic energy of its atoms. The size of helium atoms relative to their spacing is shown to scale under 1950 atmospheres of pressure. The atoms have a certain, average speed, slowed down here two trillion fold from room temperature.

Assumptions

The theory for ideal gases makes the following assumptions:

- The gas consists of very small particles known as molecules. This smallness of their size is such that the total volume of the individual gas molecules added up is negligible compared to the volume of the smallest open ball containing all the molecules. This is equivalent to stating that the average distance separating the gas particles is large compared to their size.
- These particles have the same mass.
- The number of molecules is so large that statistical treatment can be applied.
- These molecules are in constant, random, and rapid motion.
- The rapidly moving particles constantly collide among themselves and with the walls of the container. All these collisions are perfectly elastic. This means, the molecules are considered to be perfectly spherical in shape, and elastic in nature.
- Except during collisions, the interactions among molecules are negligible. (That is, they exert no forces on one another.)

This implies:

1. Relativistic effects are negligible.
2. Quantum-mechanical effects are negligible. This means that the inter-particle distance is much larger than the thermal de Broglie wavelength and the molecules are treated as classical objects.
3. Because of the above two, their dynamics can be treated classically. This means that the equations of motion of the molecules are time-reversible.
    - The average kinetic energy of the gas particles depends only on the absolute temperature of the system. The kinetic theory has its own definition of

temperature, not identical with the thermodynamic definition.

- The time during collision of molecule with the container›s wall is negligible as compared to the time between successive collisions.
- Because they have mass, the gas molecules will be affected by gravity.

More modern developments relax these assumptions and are based on the Boltzmann equation. These can accurately describe the properties of dense gases, because they include the volume of the molecules. The necessary assumptions are the absence of quantum effects, molecular chaos and small gradients in bulk properties. Expansions to higher orders in the density are known as virial expansions.

An important book on kinetic theory is that by Chapman and Cowling. An important approach to the subject is called Chapman–Enskog theory.There have been many modern developments and there is an alternative approach developed by Grad based on moment expansions. In the other limit, for extremely rarefied gases, the gradients in bulk properties are not small compared to the mean free paths. This is known as the Knudsen regime and expansions can be performed in the Knudsen number.

Properties

Pressure and Kinetic Energy

In kinetic model of gases the pressure is equal to the force exerted by the atoms hitting and rebounding from a unit area of the gas container surface. Consider a gas of N molecules, each of mass m, enclosed in a cuboid of volume $V = L^3$. When a gas molecule collides with the wall of the container perpendicular to the x axis and bounces off in the opposite direction with the same speed (an elastic collision), the change in momentum is given by:

$$\Delta p = p_{i,x} - p_{f,x} = p_{i,x} - (-p_{i,x}) = 2p_{i,x} = 2mv_x,$$

where p is the momentum, i and f indicate initial and final momentum (before and after collision), x indicates that only the x direction is being considered, and v is the speed of the particle (which is the same before and after the collision).

The particle impacts one specific side wall once every

$$\Delta t = \frac{2L}{v_x},$$

where L is the distance between opposite walls.

The force due to this particle is

$$F = \frac{\Delta p}{\Delta t} = \frac{mv_x^2}{L}.$$

The total force on the wall is

$$F = \frac{Nm\overline{v_x^2}}{L},$$

where the bar denotes an average over the N particles.

Since the motion of the particles is random and there is no bias applied in any direction, the average squared speed in each direction is identical:

$$\overline{v_x^2} = \overline{v_y^2} = \overline{v_z^2}.$$

By Pythagorean theorem in three dimensions the total squared speed v is given by

$$\overline{v^2} = \overline{v_x^2} + \overline{v_y^2} + \overline{v_z^2},$$

$$\overline{v^2} = 3\overline{v_x^2}.$$

Therefore:

$$\overline{v_x^2} = \frac{\overline{v^2}}{3},$$

and the force can be written as:

$$F = \frac{Nm\overline{v^2}}{3L}.$$

This force is exerted on an area L^2. Therefore, the pressure of the gas is

$$P = \frac{F}{L^2} = \frac{Nm\overline{v^2}}{3V},$$

$$P = \frac{F}{L^2} = \frac{Nm\overline{v^2}}{3V},$$

where $V = L^3$ is the volume of the box.

In terms of the kinetic energy of a gas K:

$$P = \frac{2}{3} \times \frac{N}{V} \times K.$$

This is a first non-trivial result of the kinetic theory because it relates pressure, a macroscopic property, to the average (translational) kinetic energy per molecule $\frac{1}{2}m\overline{v^2}$ which is a microscopic property.

Temperature and Kinetic Energy

Rewriting the above result for the pressure as $PV = \frac{Nm\overline{v^2}}{3}$ we may combine it with the ideal gas law

$$PV = Nk_BT,$$

where K_B is the Boltzmann constant and T the absolute temperature defined by the ideal gas law, to obtain

$$k_BT = \frac{m\overline{v^2}}{3},$$

which leads to simplified expression of the average kinetic energy per molecule,

$$\frac{1}{2}m\overline{v^2} = \frac{3}{2}k_BT.$$

The kinetic energy of the system is N times that of a molecule, namely $K = \frac{1}{2}Nm\overline{v^2}$. Then the temperature T takes the form

$$T = \frac{m\overline{v^2}}{3k_B}$$

which becomes

$$T = \frac{2}{3}\frac{K}{Nk_B}.$$

Eq.(**3**) is one important result of the kinetic theory: *The average molecular kinetic energy is proportional to the ideal gas law's absolute temperature.* From Eq.(**1**) and Eq.(**3**), we have

$$PV = \frac{2}{3}K.$$

Thus, the product of pressure and volume per mole is proportional to the average (translational) molecular kinetic energy.

Eq.(1) and Eq.(4) are called the "classical results", which could also be derived from statistical mechanics; for more details, see:

Since there are $3N$ degrees of freedom in a monatomic-gas system with N particles, the kinetic energy per degree of freedom per molecule is

$$\frac{K}{3N} = \frac{k_BT}{2}$$

In the kinetic energy per degree of freedom, the constant of proportionality of temperature is 1/2 times Boltzmann constant. In addition to this, the temperature will decrease when the pressure drops to a certain point. This result is related to the equipartition theorem.

As noted in the article on heat capacity, diatomic gases should have 7 degrees of freedom, but the lighter gases act as if they have only 5.

Thus the kinetic energy per kelvin (monatomic ideal gas) is:

- per mole: 12.47 J
- per molecule: 20.7 yJ = 129 μeV.

At standard temperature (273.15 K), we get:

- per mole: 3406 J
- per molecule: 5.65 zJ = 35.2 meV.....

Collisions with Container

One can calculate the number of atomic or molecular collisions with a wall of a container per unit area per unit time.

Assuming an ideal gas, a derivation results in an equation for total number of collisions per unit time per area:

$$N^* = \frac{1}{4}\frac{N}{V}v_{avg} = \frac{n}{4}\sqrt{\frac{8k_BT}{\pi m}}.$$

This quantity is also known as the "impingement rate" in vacuum physics.

Speed of Molecules

From the kinetic energy formula it can be shown that

$$v_{\mathrm{rms}} = \sqrt{\frac{3k_{\mathrm{B}}T}{m}}$$

with v in m/s, T in kelvins, and m is the molecular mass (kg). The most probable speed is 81.6% of the rms speed, and the mean speeds 92.1% (isotropic distribution of speeds).

Transport Properties

The kinetic theory of gases deals not only with gases in thermodynamic equilibrium, but also very importantly with gases not in thermodynamic equilibrium. This means considering what are known as 'transport properties', such a viscosity and thermal conductivity.

PHYSICALISM

In philosophy, physicalism is the ontological thesis that "everything is physical", that there is "nothing over and above" the physical, or that everything supervenes on the physical. Physicalism is a form of ontological monism—a "one substance" view of the nature of reality as opposed to a "two-substance" (dualism) or "many-substance" (pluralism) view. Both the definition of "physical" and the meaning of physicalism have been debated.

Physicalism is closely related to materialism. Physicalism grew out of materialism with the success of the physical sciences in explaining observed phenomena. The terms are often used interchangeably, although they are sometimes distinguished, for example on the basis of physics describing more than just matter (including energy and physical law). Common arguments against physicalism include both the philosophical zombie argument and the multiple observers argument, that the existence of a physical being may imply zero or more distinct conscious entities.

Definition of Physical

The use of "physical" in physicalism is a philosophical concept and can be distinguished from alternative definitions found in the literature (e.g. Popper defined a physical proposition to be one which can at least in theory be denied by observation). A "physical property", in this context, may be a metaphysical or logical combination of properties which are physical in the ordinary sense. It is common to express the notion of "metaphysical or logical combination of properties" using the notion of supervenience: A property A is said to supervene on a property B if any change in A necessarily implies a change in B. Since any change in a combination of properties must consist of a change in at least one component property, we see that the combination does indeed supervene on the individual properties. The point of this extension is that physicalists usually suppose the existence of various abstract concepts which are non-physical in the ordinary sense of the word; so physicalism cannot be defined in way that denies the existence of these abstractions. Also, physicalism defined in terms of supervenience does not entail that all properties in the actual world are type identical to physical properties. It is, therefore, compatible with multiple realizability.

From the notion of supervenience, we see that, assuming that mental, social, and biological properties supervene on physical properties, it follows that two hypothetical worlds cannot be identical in their physical properties but differ in their mental, social or biological properties.

Two common approaches to defining "physicalism" are the theory-based and object-based approaches. The theory-based conception of physicalism proposes that "a property is physical if and only if it either is the sort of property that physical theory tells us about or else is a property which metaphysically (or logically) supervenes on the sort of property that physical theory tells us about". Likewise, the object-based conception claims that "a property is physical if and only if: it either is the sort of property required by a complete account of the intrinsic nature of paradigmatic physical objects and their constituents or else is a property which metaphysically (or logically) supervenes on the sort of property required by a complete account of the intrinsic nature of paradigmatic physical objects and their constituents".

Physicalists have traditionally opted for a "theory-based" characterization of the physical either in terms of current physics, or a future (ideal) physics. These two theory-based conceptions of the physical represent both horns of Hempel's dilemma (named after the late philosopher of science and logical empiricist Carl Gustav Hempel): an argument against theory-based understandings of the physical. Very roughly, Hempel's dilemma is that if we define the physical by reference to current physics, then physicalism is very likely to be false, as it is very likely (by pessimistic meta-induction) that much of current physics is false. But if we instead define the physical in terms of a future (ideal) or completed physics, then physicalism is hopelessly vague or indeterminate.

While the force of Hempel's dilemma against theory-based conceptions of the physical remains contested, alternative "non-theory-based" conceptions of the physical have also been proposed. Frank Jackson (1998) for example, has argued in favour of the aforementioned "object-based" conception of the physical. An objection to this proposal, which Jackson himself noted in 1998, is that if it turns out that panpsychism or panprotopsychism is true, then such a non-materialist understanding of the physical gives the counterintuitive result that physicalism is, nevertheless, also true since such properties will figure in a complete account of paradigmatic examples of the physical.

David Papineau and Barbara Montero have advanced and subsequently defended a "via negativa" characterization of the physical. The gist of the via negativa strategy is to understand the physical in terms of what it is not: the mental. In other words, the via negativa strategy understands the physical as "the non-mental". An objection to the via negativa conception of the physical is that (like the object-based conception) it doesn't have the resources to distinguish neutral monism (or panprotopsychism) from physicalism.

Supervenience-Based Definitions of Physicalism

Adopting a supervenience-based account of the physical, the definition of physicalism as "all properties are physical" can be unravelled to:

1) Physicalism is true at a possible world w if and only if any world that is a physical duplicate of w is also a duplicate of w simpliciter.

Applied to the actual world (our world), statement 1 above is the claim that physicalism is true at the actual world if and only if at every possible world in which the physical properties and laws of the actual world are instantiated, the non-physical (in the ordinary sense of the word) properties of the actual world are instantiated as well. To borrow a metaphor from Saul Kripke (1972), the truth of physicalism at the actual world entails that once God has instantiated or "fixed" the physical properties and laws of our world, then God's work is done; the rest comes "automatically".

Unfortunately, statement 1 fails to capture even a necessary condition for physicalism to be true at a world w. To see this, imagine a world in which there are only physical properties—if physicalism is true at any world it is true at this one. But one can conceive physical duplicates of such a world that are not also duplicates simpliciter of it: worlds that have the same physical properties as our imagined one, but with some additional property or properties. A world might contain "epiphenomenal ectoplasm", some additional pure experience that does not interact with the physical components of the world and is not necessitated by them (does not supervene on them). To handle the epiphenomenal ectoplasm problem, statement 1 can be modified to include a "that's-all" or "totality" clause or be restricted to "positive" properties. Adopting the former suggestion here, we can reformulate statement 1 as follows:

2) Physicalism is true at a possible world w if and only if any world that is a minimal physical duplicate of w is a duplicate of w simpliciter.

Applied in the same way, statement 2 is the claim that physicalism is true at a possible world w if and only if any world that is a physical duplicate of w (without any further changes), is duplicate of w without qualification. This allows a world in which there are only physical properties to be counted as one at which physicalism is true, since worlds in which there is some extra stuff are not "minimal" physical duplicates of such a world, nor are they minimal physical duplicates of worlds that contain some non-physical properties that are metaphysically necessitated by the physical.

But while statement 2 overcomes the problem of worlds at which there is some extra stuff (sometimes referred to as the "epiphenomenal ectoplasm problem") it faces a different challenge: the so-called "blockers problem". Imagine a world where the relation between the physical and non-physical properties at this world (call the world w1) is slightly weaker than metaphysical necessitation, such that a certain kind of non-physical intervener—"a blocker"—could, were it to exist at w1, prevent the non-physical properties in w1 from being instantiated by the instantiation of the physical properties at w1. Since statement 2 rules out worlds which are physical duplicates of w1 that also contain non-physical interveners by virtue of the minimality, or that's-all clause, statement 2 gives the (allegedly) incorrect result that physicalism is true at w1. One response to this problem is to abandon statement 2 in favour of the alternative possibility mentioned earlier in which supervenience-based formulations of physicalism are restricted to what David Chalmers (1996) calls "positive properties". A positive property is one that "... if instantiated in a world W, is also instantiated by the corresponding individual in all worlds that contain W as a proper part." Following this suggestion, we can then formulate physicalism as follows:

3) Physicalism is true at a possible world w if and only if any world that is a physical duplicate of w is a positive duplicate of w.

On the face of it, statement 3 seems able to handle both the epiphenomenal ectoplasm problem and the blockers problem. With regard to the former, statement 3 gives the correct result that a purely physical world is one at which physicalism is true, since worlds in which there is some extra stuff are positive duplicates of a purely physical world. With regard to the latter, statement 3 appears to have the consequence that worlds in which there are blockers are worlds where positive non-physical properties of w1 will be absent, hence w1 will not be counted as a world at which physicalim is true. Daniel Stoljar (2010) objects to this response to the blockers problem on the basis that since the non-physical properties of w1 aren't instantiated at a world in which there is a blocker, they are not positive properties in Chalmers' (1996) sense, and so statement 3 will count w1 as a world at which physicalism is true after all.

A further problem for supervenience-based formulations of physicalism is the so-called "necessary beings problem". A necessary being in this context is a non-physical being that exists in all possible worlds (for example what theists refer to as God). A necessary being is compatible with all the definitions provided, because it is supervenient on everything; yet it is usually taken to contradict the notion that everything is physical. So any supervenience-based formulation of physicalism will at best state a necessary but not sufficient condition for the truth of physicalism.

Additional objections have been raised to the above definitions provided for supervenience physicalism: one could imagine an alternate world that differs only by the presence of a single ammonium molecule (or physical property), and yet based on statement 1, such a world might be completely different in terms of its distribution of mental properties. Furthermore, there are differences expressed concerning the modal status of physicalism; whether it is a necessary truth, or is only true in a world which conforms to certain conditions (i.e. those of physicalism).

Realisation Physicalism

Closely related to supervenience physicalism, is realisation physicalism, the thesis that every instantiated property is either physical or is realised by a physical property.

Token Physicalism

Token physicalism is the proposition that "for every actual particular (object, event or process) x, there is some physical particular y such that x = y". It is intended to capture the idea of "physical mechanisms". Token physicalism is compatible with property dualism, in which all substances are "physical", but physical objects may have mental properties as well as physical properties. Token physicalism is not however equivalent to supervenience physicalism. Firstly, token physicalism does not imply supervenience physicalism because the former does not rule out the possibility of non-supervenient properties (provided that they are associated only with physical particulars). Secondarily, supervenience physicalism does not imply token physicalism, for the former allows supervenient objects (such as a "nation", or "soul") that are not equal to any physical object.

Reductionism and Emergentism

Reductionism

There are multiple versions of reductionism. In the context of physicalism, the reductions referred to are of a "linguistic" nature, allowing discussions of, say, mental phenomena to be translated into discussions of physics. In one formulation, every concept is analysed in terms of a physical concept. One counter-argument to this supposes there may be an additional class of expressions which is non-physical but which increases the expressive power of a theory. Another version of reductionism is based on the requirement that one theory (mental or physical) be logically derivable from a second.

The combination of reductionism and physicalism is usually called reductive physicalism in the philosophy of mind. The opposite view is non-reductive physicalism. Reductive physicalism is the view that mental states are both nothing over and above physical states and reducible to physical states. One version of reductive physicalism is type physicalism or mind-body identity theory. Type physicalism asserts that "for every actually instantiated property F, there is some physical property G such that F=G". Unlike token physicalism, type physicalism entails supervenience physicalism.

A common argument against reductive physicalism is multiple realizability, the possibility that a psychological process (say) could be instantiated by many different neurological processes (even non-neurological processes, in the case of machine or alien intelligence). For in this case, the neurological terms translating a psychological term must be disjunctions over the possible instantiations, and it is argued that no physical law can use these disjunctions as terms. Type physicalism was the original target of the multiple realizability argument.

Emergentism

There are two versions of emergentism, the strong version and the weak version. Supervenience physicalism has been seen as a strong version of emergentism, in which the subject's psychological experience is considered genuinely novel. Non-reductive physicalism, on the other side, is a weak version of emergentism because it does not need that the subject's psychological experience be novel. The strong version of emergentism is incompatible with physicalism. Since there are novel mental states, mental states are not nothing over and above physical states. However, the weak version of emergentism is compatible with physicalism.

We can see that emergentism is actually a very broad view. Some forms of emergentism appear either incompatible with physicalism or equivalent to it (e.g. posteriori physicalism), others appear to merge both dualism and supervenience. Emergentism compatible with dualism claims that mental states and physical states are metaphysically distinct while maintaining the supervenience of mental states on physical states. This proposition however contradicts supervenience physicalism, which asserts a denial of dualism.

A Priori versus a Posteriori Physicalism

Physicalists hold that physicalism is true. A natural question for physicalists, then, is whether the truth of physicalism is deducible a priori from the nature of the physical world (i.e., the inference is justified independently of experience, even though the nature of the physical world can itself only be determined through experience) or can only be deduced a posteriori (i.e., the justification of the inference itself is dependent upon experience). So-called "a priori physicalists" hold that from knowledge of the conjunction of all physical truths, a totality or that's-all truth (to rule out non-physical epiphenomena, and enforce the closure of the physical world), and some primitive indexical truths such as "I am A" and "now is B", the truth of physicalism is knowable a priori. Let "P" stand for the conjunction of all physical truths and laws, "T" for a that's-all truth, "I" for the indexical "centering" truths, and "N" for any [presumably non-physical] truth at the actual world. We can then, using the material conditional "→", represent a priori physicalism as the thesis that PTI → N is knowable a priori. An important wrinkle here is that the concepts in N must be possessed non-deferentially in order for PTI → N to be knowable a priori. The suggestion, then, is that possession of the concepts in the consequent, plus the empirical information in the antecedent is sufficient for the consequent to be knowable a priori.

An "a posteriori physicalist", on the other hand, will reject the claim that PTI → N is knowable a priori. Rather, they would hold that the inference from PTI to N is justified by metaphysical considerations that in turn can be derived from experience. So the claim then is that "PTI and not N" is metaphysically impossible.

One commonly issued challenge to a priori physicalism and to physicalism in general is the "conceivability argument", or zombie argument. At a rough approximation, the conceivability argument runs as follows:

P1) PTI and not Q (where "Q" stands for the conjunction of all truths about consciousness, or some "generic" truth about someone being "phenomenally" conscious [i.e., there is "something it is like" to be a person x]) is conceivable (i.e., it is not knowable a priori that PTI and not Q is false).

P2) If PTI and not Q is conceivable, then PTI and not Q is metaphysically possible.

P3) If PTI and not Q is metaphysically possible then physicalism is false.

C) Physicalism is false.

Here proposition P3 is a direct application of the supervenience of consciousness, and hence of any supervenience-based version of physicalism: If PTI and not Q is possible, there is some possible world where it is true. This world differs from [the relevant indexing on] our world, where PTIQ is true. But the other world is a minimal physical duplicate of our world, because PT is true there. So there is a possible world which is a minimal physical duplicate of our world, but not a full duplicate; this contradicts the definition of physicalism that we saw above.

Since a priori physicalists hold that PTI → N is a priori, they are committed to denying P1) of the conceivability argument. The a priori physicalist, then, must argue that PTI and not Q, on ideal rational reflection, is incoherent or contradictory.

A posteriori physicalists, on the other hand, generally accept P1) but deny P2)--the move from "conceivability to metaphysical possibility". Some a posteriori physicalists think that unlike the possession of most, if not all other empirical concepts, the possession of consciousness has the special property that the presence of PTI and the absence of consciousness will be conceivable—even though, according to them, it is knowable a posteriori that PTI and not Q is not metaphysically possible. These a posteriori physicalists endorse some version of what Daniel Stoljar (2005) has called "the phenomenal concept strategy". Roughly speaking, the phenomenal concept strategy is a label for those a posteriori physicalists who attempt to show that it is only the concept of consciousness—not the property—that is in some way "special" or sui generis. Other a posteriori physicalists eschew the phenomenal concept strategy, and argue that even ordinary macroscopic truths such as "water covers 60% of the earth's surface" are not knowable a priori from PTI and a non-deferential grasp of the concepts "water" and "earth" et cetera. If this is correct, then we should (arguably) conclude that conceivability does not entail metaphysical possibility, and P2) of the conceivability argument against physicalism is false.

Other Views

Strawsonian Physicalism

Galen Strawson's realistic physicalism (or "realistic monism") entails panpsychism – or at least micropsychism. Strawson argues that "many—perhaps most—of those who call themselves physicalists or materialists [are mistakenly] committed to the thesis that physical stuff is, in itself, in its fundamental nature, something wholly and utterly non-experiential... even when they are prepared to admit with Eddington that physical stuff has, in itself, 'a nature capable of manifesting itself as mental activity', i.e. as experience or consciousness". Because experiential phenomena allegedly cannot be emergent from wholly non-experiential phenomena, philosophers are driven to substance dualism, property dualism, eliminative materialism and "all other crazy attempts at wholesale mental-to-non-mental reduction".

Real physicalists must accept that at least some ultimates are intrinsically experience-involving. They must at least embrace micropsychism. Given that everything concrete is physical, and that everything physical is constituted out of physical ultimates, and that experience is part of concrete reality, it seems the only reasonable position, more than just an 'inference to the best explanation'... Micropsychism is not yet panpsychism, for as things stand realistic physicalists can conjecture that only some types of ultimates are intrinsically experiential. But they must allow that panpsychism may be true, and the big step has already been taken with micropsychism, the admission that at least some ultimates must be experiential. 'And were the inmost essence of things laid open to us' I think that the idea that some but not all physical ultimates are experiential would look like the idea that some but not all physical ultimates are spatio-temporal (on the assumption that spacetime is indeed a fundamental feature of reality). I would bet a lot against there being such radical heterogeneity at the very bottom of things. In fact (to disagree with my earlier self) it is hard to see why this view would not count as a form of dualism... So now I can say that physicalism, i.e. real physicalism, entails panexperientialism or panpsychism. All physical stuff is energy, in one form or another, and all energy, I trow, is an experience-involving phenomenon. This sounded crazy to me for a long time, but I am quite used to it, now that I know that there is no alternative short of 'substance dualism'... Real physicalism, realistic physicalism, entails panpsychism, and whatever problems are raised by this fact are problems a real physicalist must face.

REFERENCES

1. Berryman, Sylvia, "Ancient Atomism", Stanford Encyclopedia of Philosophy (Fall 2008 Edition), Edward N. Zalta (ed.)
2. Weisstein, Eric W. "Lavoisier, Antoine (1743-1794)". scienceworld.wolfram.com. Retrieved 2009-08-01.
3. Proust, Joseph Louis. "Researches on Copper", excerpted from Ann. chim. 32, 26-54 (1799) [as translated and reproduced in Henry M. Leicester and Herbert S. Klickstein, A Source Book in Chemistry, 1400–1900 (Cambridge, Massachusetts: Harvard, 1952)]. Retrieved on August 29, 2007.
4. Andrew G. van Melsen (1952). From Atomos to Atom. Mineola, N.Y.: Dover Publications. ISBN 0-486-49584-1.
5. Dalton, John. "On the Absorption of Gases by Water and Other Liquids", in Memoirs of the Literary and Philosophical Society of Manchester. 1803. Retrieved on August 29, 2007.
6. Johnson, Chris. "Avogadro - his contribution to chemistry". Retrieved 2009-08-01.[dead link]
7. Alan J. Rocke (1984). Chemical Atomism in the Nineteenth Century. Columbus: Ohio State University Press.

8. Avogadro, Amedeo (1811). "Essay on a Manner of Determining the Relative Masses of the Elementary Molecules of Bodies, and the Proportions in Which They Enter into These Compounds". Journal de Physique. 73: 58–76.
9. Einstein, A. (1905). "Über die von der molekularkinetischen Theorie der Wärme geforderte Bewegung von in ruhenden Flüssigkeiten suspendierten Teilchen". Annalen der Physik. 322 (8): 549. Bibcode:1905AnP...322..549E. doi:10.1002/andp.19053220806.
10. Thomson, J.J. (1897). "Cathode rays" ([facsimile from Stephen Wright, Classical Scientific Papers, Physics (Mills and Boon, 1964)]). Philosophical Magazine. 44 (269): 293. doi:10.1080/14786449708621070.
11. Whittaker, E. T. (1951), A history of the theories of aether and electricity. Vol 1, Nelson, London
12. Thomson, J.J. (1904). "On the Structure of the Atom: an Investigation of the Stability and Periods of Oscillation of a number of Corpuscles arranged at equal intervals around the Circumference of a Circle; with Application of the Results to the Theory of Atomic Structure". Philosophical Magazine. 7 (39): 237. doi:10.1080/14786440409463107.
13. Geiger, H (1910). "The Scattering of the α-Particles by Matter". Proceedings of the Royal Society. A 83: 492–504.
14. Rutherford, Ernest (1911). "The Scattering of α and β Particles by Matter and the Structure of the Atom" (PDF). Philosophical Magazine. 21 (4): 669. Bibcode:2012PMag...92..379R. doi:10.1080/14786435.2011.617037.
15. Bohr, Niels (1913). "On the constitution of atoms and molecules" (PDF). Philosophical Magazine. 26 (153): 476–502. doi:10.1080/14786441308634993.
16. "Frederick Soddy, The Nobel Prize in Chemistry 1921". Nobel Foundation. Retrieved 2008-01-18.
17. Thomson, J.J. (1913). "Rays of positive electricity". Proceedings of the Royal Society. A 89 (607): 1–20. Bibcode:1913RSPSA..89....1T. doi:10.1098/rspa.1913.0057. [as excerpted in Henry A. Boorse & Lloyd Motz, The World of the Atom, Vol. 1 (New York: Basic Books, 1966)]. Retrieved on August 29, 2007.
18. Rutherford, Ernest (1919). "Collisions of alpha Particles with Light Atoms. IV. An Anomalous Effect in Nitrogen". Philosophical Magazine. 37 (222): 581. doi:10.1080/14786440608635919.
19. Chadwick, James (1932). "Possible Existence of a Neutron" (PDF). Nature. 129 (3252): 312. Bibcode:1932Natur.129Q.312C. doi:10.1038/129312a0.
20. Schrödinger, Erwin (1926). "Quantisation as an Eigenvalue Problem". Annalen der Physik. 81 (18): 109–139. Bibcode:1926AnP...386..109S. doi:10.1002/andp.19263861802.
21. Mahanti, Subodh. "Erwin Schrödinger: The Founder of Quantum Wave Mechanics". Retrieved 2009-08-01.
22. Mahanti, Subodh. "Max Born: Founder of Lattice Dynamics". Retrieved 2009-08-01.

23. Greiner, Walter. “Quantum Mechanics: An Introduction”. Retrieved 2010-06-14.
24. Heisenberg, W. (1927). “Über den anschaulichen Inhalt der quantentheoretischen Kinematik und Mechanik”. Zeitschrift für Physik (in German). 43 (3–4): 172–198. Bibcode:1927ZPhy...43..172H. doi:10.1007/BF01397280.
25. Milton Orchin; Roger Macomber; Allan Pinhas; R. Wilson. “The Vocabulary and Concepts of Organic Chemistry, Second Edition,”(PDF). Retrieved 2010-06-14.
26. Clericuzio, Antonio. Elements, Principles, and Corpuscles; a study of atomism and chemistry in the seventeenth century. Dordrecht; Boston: Kluwer Academic Publishers, 2000.
27. Cornford, Francis MacDonald. Plato›s Cosmology: The Timaeus of Plato. New York: Liberal Arts Press, 1957.
28. Dijksterhuis, E. The Mechanization of the World Picture. Trans. by C. Dikshoorn. New York: Oxford University Press, 1969. ISBN 0-691-02396-4
29. Firth, Raymond. Religion: A Humanist Interpretation. Routledge, 1996. ISBN 0-415-12897-8.
30. Gangopadhyaya, Mrinalkanti. Indian Atomism: history and sources. Atlantic Highlands, New Jersey: Humanities Press, 1981. ISBN 0-391-02177-X
31. Gardet, L. “djuz›” in Encyclopaedia of Islam CD-ROM Edition, v. 1.1. Leiden: Brill, 2001.
32. Gregory, Joshua C. A Short History of Atomism. London: A. and C. Black, Ltd, 1981.
33. Kargon, Robert Hugh. Atomism in England from Hariot to Newton. Oxford: Clarendon Press, 1966.
34. Lloyd, G. E. R. Aristotle: The Growth and Structure of his Thought. Cambridge: Cambridge University Press, 1968. ISBN 0-521-09456-9
35. Lloyd, G. E. R. Greek Science After Aristotle. New York: W. W. Norton, 1973. ISBN 0-393-00780-4
36. Marmara, Michael E. “Causation in Islamic Thought.” Dictionary of the History of Ideas. New York: Charles Scribner›s Sons, 1973-74. online at the of Virginia Electronic Text Center.
37. McEvilley, Thomas (2002). The Shape of Ancient Thought: Comparative Studies in Greek and Indian Philosophies. New York: Allworth Communications Inc. ISBN 1-58115-203-5.
38. Redondi, Pietro. Galileo Heretic. Translated by Raymond Rosenthal. Princeton, NJ: Princeton University Press, 1987. ISBN 0-691-02426-X
39. Bennett, K., and McLaughlin, B. 2011. “Supervenience.” In Stanford Encyclopedia of Philosophy, ed. E. Zalta. http://plato.stanford.edu.
40. Chalmers, D. 1996. The Conscious Mind. New York: Oxford University Press.
41. Chalmers, D.; Jackson, F. (2001). “Conceptual analysis and reductive explanation”. Philosophical Review. 110: 315–361. doi:10.1215/00318108-110-3-315.
42. Chalmers, D. 2009. “The Two-Dimensional Argument Against Materialism.” In

Oxford Handbook of Philosophy of Mind, ed. B. McLaughlin. Oxford: Oxford University Press, pp. 313–335.

43. Hawthorne, J (2002). "Blocking Definitions of Materialism". Philosophical Studies. 110: 103–113. doi:10.1023/a:1020200213934.
44. Hempel, C. 1969. "Reduction: Ontological and Linguistic Facets." In Essays in Honor of Ernest Nagel. eds. S. Morgenbesser, et al. New York: St Martin›s Press.
45. Hempel, C (1980). "Comment on Goodman›s Ways of Worldmaking.". Synthese. 45: 139–199.
46. Jackson, F. 1998. From Metaphysics to Ethics: A Defense of Conceptual Analysis. New York: Oxford University Press.
47. Judisch, N (2008). "Why ‹non-mental won›t work: On Hempel›s dilemma and the characterization of the ‹physical.›". Philosophical Studies. 140: 299–318. doi:10.1007/s11098-007-9142-8.
48. Kirk, R. (2013), The Conceptual Link from Physical to Mental, Oxford University Press, Review .
49. Kripke, S. 1972. Naming and Necessity. In Semantics of Natural Language, eds. D. Davidson and G. Harman. Dordrecht: Reidel: 253-355, 763-769.
50. Lewis, D. 1994. "Reduction of Mind." In A Companion to the Philosophy of Mind, ed. S. Guttenplan. Oxford: Blackwell, pp. 412–431.
51. Lycan, W. 2003. "Chomsky on the Mind-body Problem." In Chomsky and His Critics, eds. L. Anthony and N. Hornstein. Oxford: Blackwell
52. Melnyk, A (1997). "How To Keep The ‹Physical› in Physicalism". Journal of Philosophy. 94: 622–637. doi:10.2307/2564597.
53. Montero, B (1999). "The Body Problem". Nous. 33: 183–200. doi:10.1111/0029-4624.00149.
54. Montero, B.; Papineau, D. (2005). "A Defence of the Via Negativa Argument for Physicalism". Analysis. 65: 233–237. doi:10.1111/j.1467-8284.2005.00556.x.
55. Nagel, T (1974). "What is it like to be a bat". Philosophical Review. 4: 435–50.
56. Papineau, D. 2002. Thinking About Consciousness. Oxford: Oxford University Press.
57. Poland, J. 1994.Physicalism: The Philosophical Foundations. Oxford: Clarendon.
58. Putnam, H. 1967. "Psychological Predicates." In Art, Mind, and Religion, eds. W.H. Capitan and D.D. Merrill. Pittsburgh: University of Pittsburgh Press, pp. 37–48.
59. Smart, J.J.C. 1959. "Sensations and Brain Processes." Reprinted in Materialism and the Mind-Body Problem, ed. D. Rosenthal. Indianapolis: Hackett, 1987.
60. Smart, J.J.C. (1978). "The Content of Physicalism". Philosophical Quarterly. 28: 239–41.
61. Stoljar, D (2005). "Physicalism and Phenomenal Concepts.". Mind and Language. 20: 469–494. doi:10.1111/j.0268-1064.2005.00296.x.

62. Stoljar, D. 2009. "Physicalism." in Stanford Encyclopedia of Philosophy, ed. E. Zalta. http://plato.stanford.edu.
63. Stoljar, D. 2010. Physicalism. New York: Routledge.
64. Tye, M. 2009. Consciousness Revisited: Materialism Without Phenomenal Concepts.Cambridge Mass: MIT Press.
65. Vincente, A (2011). "Current Physics and ‹the Physical,›". British Journal of Philosophy of Science. 62: 393–416.
66. Wilson, J (2006). "On Characterizing the Physical". Philosophical Studies. 131: 69–99. doi:10.1007/s11098-006-5984-8.
67. Clausius, R. (1857), "Ueber die Art der Bewegung, welche wir Wärme nennen", Annalen der Physik, 176 (3): 353–379, Bibcode:1857AnP...176..353C, doi:10.1002/andp.18571760302
68. de Groot, S. R., W. A. van Leeuwen and Ch. G. van Weert (1980), Relativistic Kinetic Theory, North-Holland, Amsterdam.
69. Einstein, A. (1905), "Über die von der molekularkinetischen Theorie der Wärme geforderte Bewegung von in ruhenden Flüssigkeiten suspendierten Teilchen" (PDF), Annalen der Physik, 17 (8): 549–560, Bibcode:1905AnP...322..549E, doi:10.1002/andp.19053220806
70. Grad, Harold (1949), "On the Kinetic Theory of Rarefied Gases.", Communications on Pure and Applied Mathematics, 2 (4): 331–407, doi:10.1002/cpa.3160020403
71. Herapath, J. (1816), "On the physical properties of gases", Annals of Philosophy, Robert Baldwin: 56–60
72. Herapath, J. (1821), "On the Causes, Laws and Phenomena of Heat, Gases, Gravitation", Annals of Philosophy, Baldwin, Cradock, and Joy, 9: 273–293
73. Krönig, A. (1856), "Grundzüge einer Theorie der Gase", Annalen der Physik, 99 (10): 315–322, Bibcode:1856AnP...175..315K, doi:10.1002/andp.18561751008
74. Le Sage, G.-L. (1818), "Physique Mécanique des Georges-Louis Le Sage", in Prévost, Pierre, Deux Traites de Physique Mécanique, Geneva & Paris: J.J. Paschoud, pp. 1–186
75. Liboff, R. L. (1990), Kinetic Theory, Prentice-Hall, Englewood Cliffs, N. J.
76. Lomonosov, M. (1970) [1758], "On the Relation of the Amount of Material and Weight", in Henry M. Leicester, Mikhail Vasil›evich Lomonosov on the Corpuscular Theory, Cambridge: Harvard University Press, pp. 224–233
77. Mahon, Basil (2003), The Man Who Changed Everything – the Life of James Clerk Maxwell, Hoboken, New Jersey: Wiley, ISBN 0-470-86171-1
78. Maxwell, James Clerk (1873), "Molecules", Nature, 417 (6892): 903, Bibcode:2002Natur.417..903M, doi:10.1038/417903a, PMID 12087385, archived from the original (– Scholar search) on February 9, 2007
79. Smoluchowski, M. (1906), "Zur kinetischen Theorie der Brownschen Molekularbewegung und der Suspensionen", Annalen der Physik, 21 (14): 756–780, Bibcode:1906AnP...326..756V, doi:10.1002/andp.19063261405

80. Waterston, John James (1843), Thoughts on the Mental Functions (reprinted in his Papers, 3, 167, 183.)

81. Williams, M. M. R. (1971), Mathematical Methods in Particle Transport Theory, Butterworths, London.

CHAPTER 3

Matter

Chapter Contents

- **Introduction**
- **Comparison With Mass**
- **Definition**
- **Structure**
- **Phases**
- **Antimatter**
- **Other Types**
- **Historical Development**
- **Summary**

INTRODUCTION

In the classical physics observed in everyday life, if something has mass and takes up space, it is said to be composed of matter; this includes atoms (and thus molecules) and anything made up of these, but not other energy phenomena or waves such as light or sound. More generally, however, in (modern) physics, matter is not a fundamental concept because a universal definition of it is elusive: elementary constituents of atoms may not take up space individually, and massless particles may be composed to form objects that have mass (even when at rest).

All the everyday objects that we can bump into, touch or squeeze are ultimately composed of atoms. This ordinary atomic matter is in turn made up of interacting subatomic particles—usually a nucleus of protons and neutrons, and a cloud of orbiting electrons. Typically, science considers these composite particles matter because they have both rest mass and volume. By contrast, massless particles, such as photons, are not considered matter, because they have neither rest mass nor volume. However, not all particles with rest mass have a classical volume, since fundamental particles such as quarks and leptons (sometimes equated with matter) are considered "point particles" with no effective size or volume. Nevertheless, quarks and leptons together make up "ordinary matter", and their interactions contribute to the effective volume of the composite particles that make up ordinary matter.

Matter exists in states (or phases): the classical solid, liquid, and gas; as well as the more exotic plasma, Bose–Einstein condensates, fermionic condensates, and quark–gluon plasma.

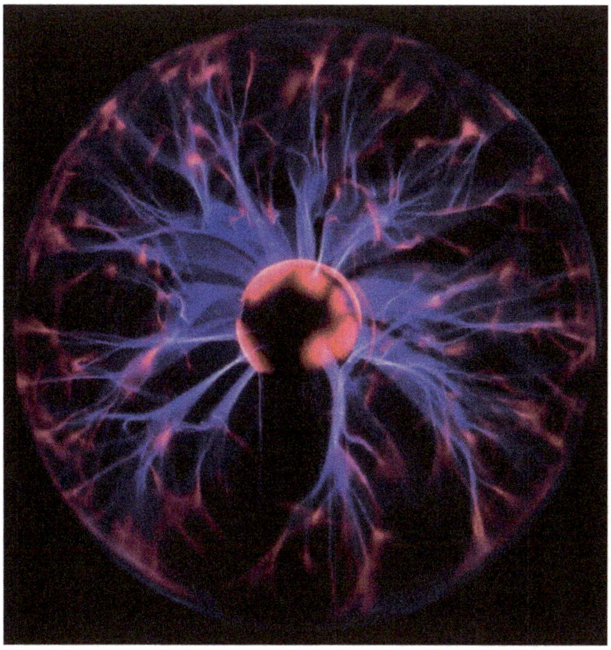

Matter is usually classified into three classical states, with plasma sometimes added as a fourth state. From top to bottom: quartz (solid), water (liquid), nitrogen dioxide (gas), and a plasma globe (plasma).

For much of the history of the natural sciences people have contemplated the exact nature of matter. The idea that matter was built of discrete building blocks, the so-called particulate theory of matter, was first put forward by the Greek philosophers Leucippus (~490 BC) and Democritus (~470–380 BC).

COMPARISON WITH MASS

Matter should not be confused with mass, as the two are not quite the same in modern physics. For example, mass is a conserved quantity, which means that its value is unchanging through time, within closed systems. However, matter is not conserved in such systems, although this is not obvious in ordinary conditions on Earth, where matter is approximately conserved. Still, special relativity shows that matter may disappear by conversion into energy, even inside closed systems, and it can also be created from energy, within such systems. However, because mass (like energy) can neither be created nor destroyed, the quantity of mass and the quantity of energy remain the same during a transformation of matter (which represents a certain amount of energy) into non-material (i.e., non-matter) energy. This is also true in the reverse transformation of energy into matter.

Different fields of science use the term matter in different, and sometimes incompatible, ways. Some of these ways are based on loose historical meanings, from a time when there was no reason to distinguish mass and matter. As such, there is no single universally agreed scientific meaning of the word "matter". Scientifically, the term "mass" is well-defined, but "matter" is not. Sometimes in the field of physics "matter" is simply equated with particles that exhibit rest mass (i.e., that cannot travel at the speed of light), such as quarks and leptons. However, in both physics and chemistry, matter exhibits both wave-like and particle-like properties, the so-called wave–particle duality.

DEFINITION

Based on Mass, Volume, and Space

The common definition of matter is anything that has mass and volume (occupies space). For example, a car would be said to be made of matter, as it has mass and volume (occupies space).

The observation that matter occupies space goes back to antiquity. However, an explanation for why matter occupies space is recent, and is argued to be a result of the phenomenon described in the Pauli Exclusion Principle. Two particular examples where the exclusion principle clearly relates matter to the occupation of space are white dwarf stars and neutron stars, discussed further below.

Based on Atoms

A definition of "matter" based on its physical and chemical structure is: matter is made up of atoms. As an example, deoxyribonucleic acid molecules (DNA) are matter under this definition because they are made of atoms. This definition can extend to include charged atoms and molecules, so as to include plasmas (gases of ions) and electrolytes (ionic solutions), which are not obviously included in the atoms definition. Alternatively, one can adopt the protons, neutrons, and electrons definition.

Based on Protons, Neutrons and Electrons

A definition of "matter" more fine-scale than the atoms and molecules definition is: matter is made up of what atoms and molecules are made of, meaning anything made of positively charged protons, neutral neutrons, and negatively charged electrons. This definition goes beyond atoms and molecules, however, to include substances made from these building blocks that are not simply atoms or molecules, for example white dwarf matter—typically, carbon and oxygen nuclei in a sea of degenerate electrons. At a microscopic level, the constituent "particles" of matter such as protons, neutrons, and electrons obey the laws of quantum mechanics and exhibit wave–particle duality. At an even deeper level, protons and neutrons are made up of quarks and the force fields (gluons) that bind them together (see Quarks and leptons definition below).

Based on Quarks and Leptons

As seen in the above discussion, many early definitions of what can be called ordinary matter were based upon its structure or building blocks. On the scale of elementary particles, a definition that follows this tradition can be stated as: ordinary matter is everything that is composed of elementary fermions, namely quarks and leptons. The connection between these formulations follows.

Leptons (the most famous being the electron), and quarks (of which baryons, such as protons and neutrons, are made) combine to form atoms, which in turn form molecules. Because atoms and molecules are said to be matter, it is natural to phrase the definition as: ordinary matter is anything that is made of the same things that atoms and molecules are made of. (However, notice that one also can make from these building blocks matter that is not atoms or molecules.) Then, because electrons are leptons, and protons, and neutrons are made of quarks, this definition in turn leads to the definition of matter as being quarks and leptons, which are the two types of elementary fermions. Carithers and Grannis state: Ordinary matter is composed entirely of first-generation particles, namely the and quarks, plus the electron and its neutrino. (Higher generations particles quickly decay into first-generation particles, and thus are not commonly encountered.)

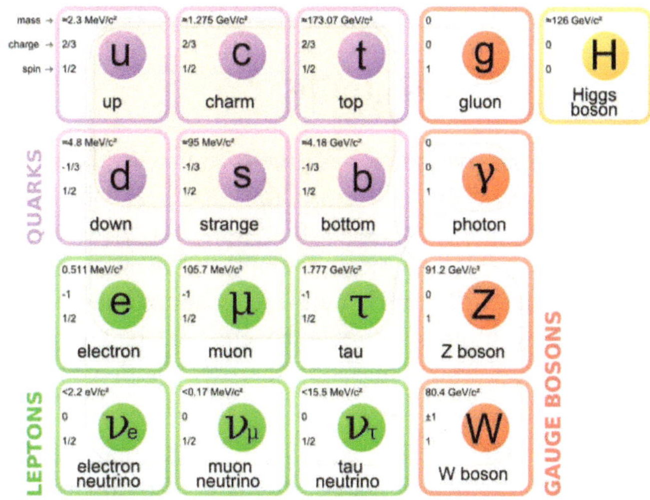

Under the "quarks and leptons" definition, the elementary and composite particles made of the quarks (in purple) and leptons (in green) would be matter—while the gauge bosons (in red) would not be matter. However, interaction energy inherent to composite particles (for example, gluons involved in neutrons and protons) contribute to the mass of ordinary matter.

This definition of ordinary matter is more subtle than it first appears. All the particles that make up ordinary matter (leptons and quarks) are elementary fermions, while all the force carriers are elementary bosons. The W and Z bosons that mediate the weak force are not made of quarks or leptons, and so are not ordinary matter, even if they have mass. In other words, mass is not something that is exclusive to ordinary matter.

The quark–lepton definition of ordinary matter, however, identifies not only the elementary building blocks of matter, but also includes composites made from the constituents (atoms and molecules, for example). Such composites contain an interaction energy that holds the constituents together, and may constitute the bulk of the mass of the composite. As an example, to a great extent, the mass of an atom is simply the sum of the masses of its constituent protons, neutrons and electrons. However, digging deeper, the protons and neutrons are made up of quarks bound together by gluon fields (see dynamics of quantum chromodynamics) and these gluons fields contribute significantly to the mass of hadrons. In other words, most of what composes the "mass" of ordinary matter is due to the binding energy of quarks within protons and neutrons. For example, the sum of the mass of the three quarks in a nucleon is approximately 12.5 MeV/c2, which is low compared to the mass of a nucleon (approximately 938 MeV/c2). The bottom line is that most of the mass of everyday objects comes from the interaction energy of its elementary components.

The Standard Model groups matter particles into three generations, where each generation consists of two quarks and two leptons. The first generation is the up and down quarks, the electron and the electron neutrino; the second includes the charm and strange quarks, the muon and the muon neutrino; the third generation consists of the top and bottom quarks and the tau and tau neutrino. The most natural explanation for this would be that quarks and leptons of higher generations are excited states of the first generations. If this turns out to be the case, it would imply that quarks and leptons are composite particles, rather than elementary particles.

Based on Theories of Relativity

In the context of relativity, mass is not an additive quantity, in the sense that one can add the rest masses of particles in a system to get the total rest mass of the system. Thus, in relativity usually a more general view is that it is not the sum of rest masses, but the energy–momentum tensor that quantifies the amount of matter. This tensor gives the rest mass for the entire system. "Matter" therefore is sometimes considered as anything that contributes to the energy–momentum of a system, that is, anything that is not purely gravity. This view is commonly held in fields that deal with general relativity such as cosmology. In this view, light and other massless particles and fields are part of matter.

The reason for this is that in this definition, electromagnetic radiation (such as light) as well as the energy of electromagnetic fields contributes to the mass of systems, and therefore appears to add matter to them. For example, light radiation (or thermal radiation) trapped inside a box would contribute to the mass of the box, as would any kind of energy inside the box, including the kinetic energy of particles held by the box. Nevertheless, isolated individual particles of light (photons) and the isolated kinetic energy of massive particles, are normally not considered to be matter.

A difference between matter and mass therefore may seem to arise when single particles are examined. In such cases, the mass of single photons is zero. For particles with rest mass, such as leptons and quarks, isolation of the particle in a frame where it is not moving, removes its kinetic energy.

A source of definition difficulty in relativity arises from two definitions of mass in common use, one of which is formally equivalent to total energy (and is thus observer dependent), and the other of which is referred to as rest mass or invariant mass and is independent of the observer. Only "rest mass" is loosely equated with matter (since it can be weighed). Invariant mass is usually applied in physics to unbound systems of particles. However, energies which contribute to the "invariant mass" may be weighed also in special circumstances, such as when a system that has invariant mass is confined and has no net momentum (as in the box example above). Thus, a photon with no mass may (confusingly) still add mass to a system in which it is trapped. The same is true of the kinetic energy of particles, which by definition is not part of their rest mass, but which does add rest mass to systems in which these particles reside (an example is the mass added by the motion of gas molecules of a bottle of gas, or by the thermal energy of any hot object).

Since such mass (kinetic energies of particles, the energy of trapped electromagnetic radiation and stored potential energy of repulsive fields) is measured as part of the mass of ordinary matter in complex systems, the "matter" status of "massless particles" and fields of force becomes unclear in such systems. These problems contribute to the lack of a rigorous definition of matter in science, although mass is easier to define as the total stress–energy above (this is also what is weighed on a scale, and what is the source of gravity).

STRUCTURE

In particle physics, fermions are particles that obey Fermi–Dirac statistics. Fermions can be elementary, like the electron—or composite, like the proton and neutron. In the Standard Model, there are two types of elementary fermions: quarks and leptons, which are discussed next.

Quarks

Quarks are particles of spin-$^{1/2}$, implying that they are fermions. They carry an electric charge of $^{-1/3}$ e (down-type quarks) or $^{+2/3}$ e (up-type quarks). For comparison, an electron has a charge of $^{-1}$ e. They also carry colour charge, which is the equivalent of the electric charge for the strong interaction. Quarks also undergo radioactive decay, meaning that they are subject to the weak interaction. Quarks are massive particles, and therefore are also subject to gravity.

Quark Properties

name	symbol	spin	electric charge (e)	mass (MeV/c^2)	mass comparable to	antiparticle	antiparticle symbol
up-type quarks							
up	u	½	+⅔	1.5 to 3.3	~ 5 electrons	antiup	$\bar{u}$
charm	c	½	+⅔	1160 to 1340	~1 proton	anticharm	$\bar{c}$
top	t	½	+⅔	169,100 to 173,300	~180 protons or ~1 tungsten atom	antitop	$\bar{t}$
down-type quarks							
down	d	½	-⅓	3.5 to 6.0	~10 electrons	antidown	$\bar{d}$
strange	s	½	-⅓	70 to 130	~ 200 electrons	antistrange	$\bar{s}$
bottom	b	½	-⅓	4130 to 4370	~ 5 protons	antibottom	$\bar{b}$

Baryonic Matter

Baryons are strongly interacting fermions, and so are subject to Fermi–Dirac statistics. Amongst the baryons are the protons and neutrons, which occur in atomic nuclei, but many other unstable baryons exist as well. The term baryon usually refers to triquarks—particles made of three quarks. "Exotic" baryons made of four quarks and one antiquark are known as the pentaquarks, but their existence is not generally accepted.

Quark structure of a proton: 2 up quarks and 1 down quark.

Baryonic matter is the part of the universe that is made of baryons (including all atoms). This part of the universe does not include dark energy, dark matter, black holes or various forms of degenerate matter, such as compose white dwarf stars and neutron stars. Microwave light seen by Wilkinson Microwave Anisotropy Probe (WMAP), suggests that only about 4.6% of that part of the universe within range of the best telescopes (that is, matter that may be visible because light could reach us from it), is made of baryonic matter. About 23% is dark matter, and about 72% is dark energy.

As a matter of fact, the great majority of ordinary matter in the universe is unseen, since visible stars and gas inside galaxies and clusters account for less than 10 per cent of the ordinary matter contribution to the mass-energy density of the universe.

Degenerate Matter

A comparison between the white dwarf IK Pegasi B (center), its A-class companion IK Pegasi A (left) and the Sun (right). This white dwarf has a surface temperature of 35,500 K.

In physics, degenerate matter refers to the ground state of a gas of fermions at a temperature near absolute zero. The Pauli exclusion principle requires that only two fermions can occupy a quantum state, one spin-up and the other spin-down. Hence, at zero temperature, the fermions fill up sufficient levels to accommodate all the available fermions—and in the case of many fermions, the maximum kinetic energy (called the Fermi energy) and the pressure of the gas becomes very large, and depends on the number of fermions rather than the temperature, unlike normal states of matter.

Degenerate matter is thought to occur during the evolution of heavy stars. The demonstration by Subrahmanyan Chandrasekhar that white dwarf stars have a maximum allowed mass because of the exclusion principle caused a revolution in the theory of star evolution.

Degenerate matter includes the part of the universe that is made up of neutron stars and white dwarfs.

Strange Matter

Strange matter is a particular form of quark matter, usually thought of as a liquid of up, down, and strange quarks. It is contrasted with nuclear matter, which is a liquid of neutrons and protons (which themselves are built out of up and down quarks), and with non-strange quark matter, which is a quark liquid that contains only up and down quarks. At high enough density, strange matter is expected to be color superconducting. Strange matter is hypothesized to occur in the core of neutron stars, or, more speculatively, as isolated droplets that may vary in size from femtometers (strangelets) to kilometers (quark stars).

Two Meanings of the Term "Strange Matter"

In particle physics and astrophysics, the term is used in two ways, one broader and the other more specific.

- The broader meaning is just quark matter that contains three flavors of quarks: up, down, and strange. In this definition, there is a critical pressure and an associated critical density, and when nuclear matter (made of protons and neutrons) is compressed beyond this density, the protons and neutrons dissociate into quarks, yielding quark matter (probably strange matter).
- The narrower meaning is quark matter that is more stable than nuclear matter. The idea that this could happen is the "strange matter hypothesis" of Bodmer and Witten. In this definition, the critical pressure is zero: the true ground state of matter is always quark matter. The nuclei that we see in the matter around us, which are droplets of nuclear matter, are actually metastable, and given enough time (or the right external stimulus) would decay into droplets of strange matter, i.e. strangelets.

Leptons

Leptons are particles of spin-1⁄2, meaning that they are fermions. They carry an electric charge of −1 e (charged leptons) or 0 e (neutrinos). Unlike quarks, leptons do not carry colour charge, meaning that they do not experience the strong interaction. Leptons also undergo radioactive decay, meaning that they are subject to the weak interaction. Leptons are massive particles, therefore are subject to gravity.

Lepton Properties

name	symbol	spin	electric charge (e)	mass (MeV/c^2)	mass comparable to	antiparticle	antiparticle symbol
charged leptons[38]							
electron	e^-	½	−1	0.5110	1 electron	antielectron	e^+
muon	μ^-	½	−1	105.7	~ 200 electrons	antimuon	μ^+
tau	τ^-	½	−1	1,777	~ 2 protons	antitau	τ^+
neutrinos[39]							
electron neutrino	ν_e	½	0	< 0.000460	< $^1/_{1000}$ electron	electron antineutrino	$\bar{\nu}_e$
muon neutrino	ν_μ	½	0	< 0.19	< ½ electron	muon antineutrino	$\bar{\nu}_\mu$
tau neutrino	ν_τ	½	0	< 18.2	< 40 electrons	tau antineutrino	$\bar{\nu}_\tau$

PHASES

In bulk, matter can exist in several different forms, or states of aggregation, known as phases, depending on ambient pressure, temperature and volume. A phase is a form of matter that has a relatively uniform chemical composition and physical properties (such as density, specific heat, refractive index, and so forth). These phases include the three familiar ones (solids, liquids, and gases), as well as more exotic states of matter (such as plasmas, superfluids, supersolids, Bose–Einstein condensates, ...). A fluid may be a liquid, gas or plasma. There are also paramagnetic and ferromagnetic phases of magnetic materials. As conditions change, matter may change from one phase into another. These phenomena are called phase transitions, and are studied in the field of thermodynamics. In nanomaterials, the vastly increased ratio of surface area to volume results in matter that can exhibit properties entirely different from those of bulk material, and not well described by any bulk phase (see nanomaterials for more details).

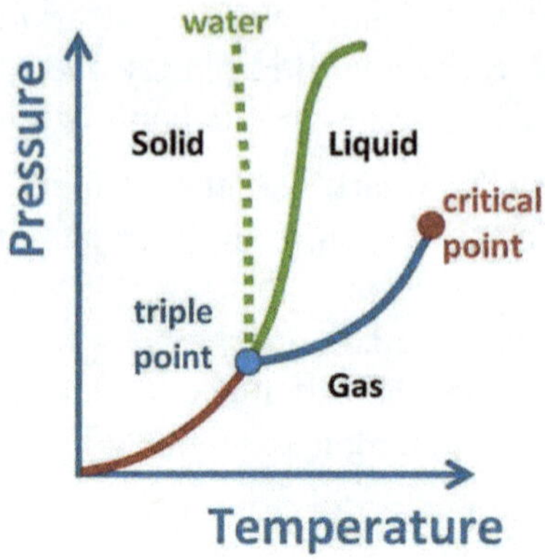

Phase diagram for a typical substance at a fixed volume. Vertical axis is Pressure, horizontal axis is Temperature. The green line marks the freezing point (above the green line is solid, below it is liquid) and the blue line the boiling point (above it is liquid and below it is gas). So, for example, at higher T, a higher P is necessary to maintain the substance in liquid phase. At the triple point the three phases; liquid, gas and solid; can coexist. Above the critical point there is no detectable difference between the phases. The dotted line shows the anomalous behavior of water: ice melts at constant temperature with increasing pressure.

Phases are sometimes called states of matter, but this term can lead to confusion with thermodynamic states. For example, two gases maintained at different pressures are in different thermodynamic states (different pressures), but in the same phase (both are gases).

ANTIMATTER

In particle physics and quantum chemistry, antimatter is matter that is composed of the antiparticles of those that constitute ordinary matter. If a particle and its antiparticle come into contact with each other, the two annihilate; that is, they may both be converted into other particles with equal energy in accordance with Einstein's equation E = mc2. These new particles may be high-energy photons (gamma rays) or other particle–antiparticle pairs. The resulting particles are endowed with an amount of kinetic energy equal to the difference between the rest mass of the products of the annihilation and the rest mass of the original particle–antiparticle pair, which is often quite large.

Antimatter is not found naturally on Earth, except very briefly and in vanishingly small quantities (as the result of radioactive decay, lightning or cosmic rays). This is because antimatter that came to exist on Earth outside the confines of a suitable physics laboratory would almost instantly meet the ordinary matter that Earth is made of, and be annihilated. Antiparticles and some stable antimatter (such as antihydrogen) can be made in tiny amounts, but not in enough quantity to do more than test a few of its theoretical properties.

There is considerable speculation both in science and science fiction as to why the observable universe is apparently almost entirely matter, and whether other places are almost entirely antimatter instead. In the early universe, it is thought that matter and antimatter were equally represented, and the disappearance of antimatter requires an asymmetry in physical laws called the charge parity (or CP symmetry) violation. CP symmetry violation can be obtained from the Standard Model, but at this time the apparent asymmetry of matter and antimatter in the visible universe is one of the great unsolved problems in physics. Possible processes by which it came about are explored in more detail under baryogenesis.

OTHER TYPES

Ordinary matter, in the quarks and leptons definition, constitutes about 4% of the energy of the observable universe. The remaining energy is theorized to be due to exotic forms, of which 23% is dark matter and 73% is dark energy.

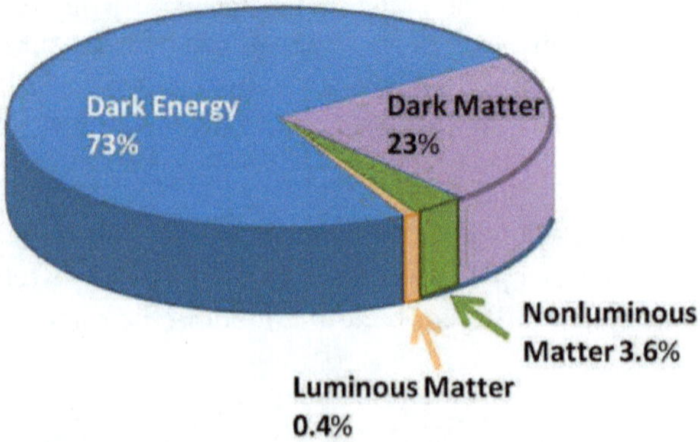

Pie chart showing the fractions of energy in the universe contributed by different sources. Ordinary matter is divided into luminous matter (the stars and luminous gases and 0.005% radiation) and nonluminous matter (intergalactic gas and about 0.1% neutrinos and 0.04% supermassive black holes). Ordinary matter is uncommon. Modeled after Ostriker and Steinhardt. For more information, see NASA.

Dark Matter

In astrophysics and cosmology, dark matter is matter of unknown composition that does not emit or reflect enough electromagnetic radiation to be observed directly, but whose presence can be inferred from gravitational effects on visible matter. Observational evidence of the early universe and the big bang theory require that this matter have energy and mass, but is not composed of either elementary fermions (as above) OR gauge bosons. The commonly accepted view is that most of the dark matter is non-baryonic in nature. As such, it is composed of particles as yet unobserved in the laboratory. Perhaps they are supersymmetric particles, which are not Standard Model particles, but relics formed at very high energies in the early phase of the universe and still floating about.

Dark Energy

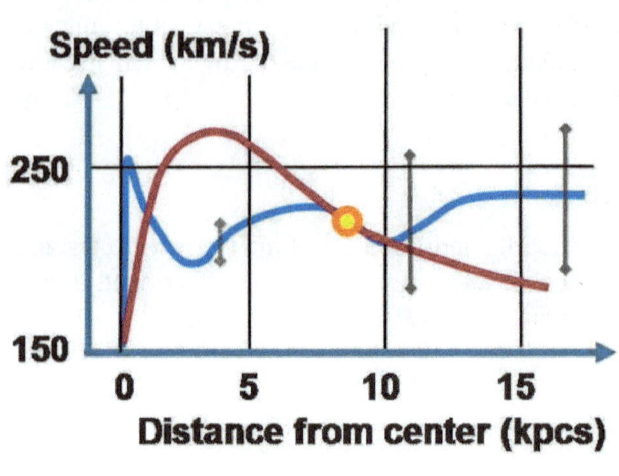

Galaxy rotation curve for the Milky Way. Vertical axis is speed of rotation about the galactic center. Horizontal axis is distance from the galactic center. The sun is marked with a yellow ball. The observed curve of speed of rotation is blue. The predicted curve based upon stellar mass and gas in the Milky Way is red. The difference is due to dark matter or perhaps a modification of the law of gravity. Scatter in observations is indicated roughly by gray bars.

In cosmology, dark energy is the name given to the antigravitating influence that is accelerating the rate of expansion of the universe. It is known not to be composed of known particles like protons, neutrons or electrons, nor of the particles of dark matter, because these all gravitate.

Fully 70% of the matter density in the universe appears to be in the form of dark energy. Twenty-six percent is dark matter. Only 4% is ordinary matter. So less than 1 part in 20 is made out of matter we have observed experimentally or described in the standard model of particle physics. Of the other 96%, apart from the properties just mentioned, we know absolutely nothing.

Exotic Matter

Exotic matter is a hypothetical concept of particle physics. It covers any material that violates one or more classical conditions or is not made of known baryonic particles. Such materials would possess qualities like negative mass or being repelled rather than attracted by gravity.

HISTORICAL DEVELOPMENT

Antiquity (c. 610 BC–c. 322 BC)

The pre-Socratics were among the first recorded speculators about the underlying nature of the visible world. Thales (c. 624 BC–c. 546 BC) regarded water as the fundamental material of the world. Anaximander (c. 610 BC–c. 546 BC) posited that the basic material was wholly characterless or limitless: the Infinite (apeiron). Anaximenes (flourished 585 BC, d. 528 BC) posited that the basic stuff was pneuma or air. Heraclitus (c. 535–c. 475 BC) seems to say the basic element is fire, though perhaps he means that all is change. Empedocles (c. 490–430 BC) spoke of four elements of which everything was made: earth, water, air, and fire. Meanwhile, Parmenides argued that change does not exist, and Democritus argued that everything is composed of minuscule, inert bodies of all shapes called atoms, a philosophy called atomism. All of these notions had deep philosophical problems.

Aristotle (384 BC – 322 BC) was the first to put the conception on a sound philosophical basis, which he did in his natural philosophy, especially in Physics book I. He adopted as reasonable suppositions the four Empedoclean elements, but added a fifth, aether. Nevertheless, these elements are not basic in Aristotle's mind. Rather they, like everything else in the visible world, are composed of the basic principles matter and form.

For my definition of matter is just this—the primary substratum of each thing, from which it comes to be without qualification, and which persists in the result.

—Aristotle, Physics I:9:192a32

The word Aristotle uses for matter, λη (hyle or hule), can be literally translated as wood or timber, that is, "raw material" for building. Indeed, Aristotle›s conception of matter is intrinsically linked to something being made or composed. In other words, in contrast to the early modern conception of matter as simply occupying space, matter for Aristotle is definitionally linked to process or change: matter is what underlies a change of substance. For example, a horse eats grass: the horse changes the grass into itself; the grass as such does not persist in the horse, but some aspect of it—its matter—does. The matter is not specifically described (e.g., as atoms), but consists of whatever persists in the change of substance from grass to horse. Matter in this understanding does not exist independently (i.e., as a substance), but exists interdependently (i.e., as a "principle") with form and only insofar as it underlies change. It can be helpful to conceive of the relationship of matter and form as very similar to that between parts and whole. For Aristotle, matter as such can only receive actuality from form; it has no activity or actuality in itself, similar to the way that parts as such only have their existence in a whole (otherwise they would be independent wholes).

Seventeenth and Eighteenth Centuries

René Descartes (1596–1650) originated the modern conception of matter. He was primarily a geometer. Instead of, like Aristotle, deducing the existence of matter from the physical reality of change, Descartes arbitrarily postulated matter to be an abstract, mathematical substance that occupies space:

So, extension in length, breadth, and depth, constitutes the nature of bodily substance; and thought constitutes the nature of thinking substance. And everything else attributable to body presupposes extension, and is only a mode of extended

—René Descartes, Principles of Philosophy

For Descartes, matter has only the property of extension, so its only activity aside from locomotion is to exclude other bodies: this is the mechanical philosophy. Descartes makes an absolute distinction between mind, which he defines as unextended, thinking substance, and matter, which he defines as unthinking, extended substance. They are independent things. In contrast, Aristotle defines matter and the formal/forming principle as complementary principles that together compose one independent thing (substance). In short, Aristotle defines matter (roughly speaking) as what things are actually made of (with a potential independent existence), but Descartes elevates matter to an actual independent thing in itself.

The continuity and difference between Descartes' and Aristotle's conceptions is noteworthy. In both conceptions, matter is passive or inert. In the respective conceptions matter has different relationships to intelligence. For Aristotle, matter and intelligence (form) exist together in an interdependent relationship, whereas for Descartes, matter and intelligence (mind) are definitionally opposed, independent substances.

Descartes' justification for restricting the inherent qualities of matter to extension is its permanence, but his real criterion is not permanence (which equally applied to color and resistance), but his desire to use geometry to explain all material properties. Like Descartes, Hobbes, Boyle, and Locke argued that the inherent properties of bodies were limited to extension, and that so-called secondary qualities, like color, were only products of human perception.

Isaac Newton (1643–1727) inherited Descartes' mechanical conception of matter. In the third of his "Rules of Reasoning in Philosophy", Newton lists the universal qualities of matter as "extension, hardness, impenetrability, mobility, and inertia". Similarly in Optics he conjectures that God created matter as "solid, massy, hard, impenetrable, movable particles", which were "...even so very hard as never to wear or break in pieces". The "primary" properties of matter were amenable to mathematical description, unlike "secondary" qualities such as color or taste. Like Descartes, Newton rejected the essential nature of secondary qualities.

Newton developed Descartes' notion of matter by restoring to matter intrinsic properties in addition to extension (at least on a limited basis), such as mass. Newton's use of gravitational force, which worked "at a distance", effectively repudiated Descartes' mechanics, in which interactions happened exclusively by contact.

Though Newton's gravity would seem to be a power of bodies, Newton himself did not admit it to be an essential property of matter. Carrying the logic forward more consistently, Joseph Priestley (1733-1804) argued that corporeal properties transcend contact mechanics: chemical properties require the capacity for attraction. He argued matter has other inherent powers besides the so-called primary qualities of Descartes, et al.

Nineteenth and Twentieth Centuries

Since Priestley's time, there has been a massive expansion in knowledge of the constituents of the material world (viz., molecules, atoms, subatomic particles), but there has been no further development in the definition of matter. Rather the question has been set aside. Noam Chomsky (born 1928) summarizes the situation that has prevailed since that time:

What is the concept of body that finally emerged? The answer is that there is no clear and definite conception of body Rather, the material world is whatever we discover it to be, with whatever properties it must be assumed to have for the purposes of explanatory theory. Any intelligible theory that offers genuine explanations and that can be assimilated to the core notions of physics becomes part of the theory of the material world, part of our account of body. If we have such a theory in some domain, we seek to assimilate it to the core notions of physics, perhaps modifying these notions as we carry out this enterprise.

—Noam Chomsky, Language and problems of knowledge: the Managua lectures, p. 144

So matter is whatever physics studies and the object of study of physics is matter: there is no independent general definition of matter, apart from its fitting into the methodology of measurement and controlled experimentation. In sum, the boundaries between what constitutes matter and everything else remains as vague as the demarcation problem of delimiting science from everything else.

In the 19th century, following the development of the periodic table, and of atomic theory, atoms were seen as being the fundamental constituents of matter; atoms formed molecules and compounds.

The common definition in terms of occupying space and having mass is in contrast with most physical and chemical definitions of matter, which rely instead upon its structure and upon attributes not necessarily related to volume and mass. At the turn of the nineteenth century, the knowledge of matter began a rapid evolution.

Aspects of the Newtonian view still held sway. James Clerk Maxwell discussed matter in his work Matter and Motion. He carefully separates "matter" from space and time, and defines it in terms of the object referred to in Newton's first law of motion.

However, the Newtonian picture was not the whole story. In the 19th century, the term "matter" was actively discussed by a host of scientists and philosophers, and a brief outline can be found in Levere. A textbook discussion from 1870 suggests matter is what is made up of atoms:

Three divisions of matter are recognized in science: masses, molecules and atoms.

A Mass of matter is any portion of matter appreciable by the senses.

A Molecule is the smallest particle of matter into which a body can be divided without losing its identity.

An Atom is a still smaller particle produced by division of a molecule.

Rather than simply having the attributes of mass and occupying space, matter was held to have chemical and electrical properties. In 1909 the famous physicist J. J. Thomson (1856-1940) wrote about the "constitution of matter" and was concerned with the possible connection between matter and electrical charge.

There is an entire literature concerning the "structure of matter", ranging from the "electrical structure" in the early 20th century, to the more recent "quark structure of matter", introduced today with the remark: Understanding the quark structure of matter has been one of the most important advances in contemporary physics. In this connection, physicists speak of matter fields, and speak of particles as "quantum excitations of a mode of the matter field". And here is a quote from de Sabbata and Gasperini: "With the word "matter" we denote, in this context, the sources of the interactions, that is spinor fields (like quarks and leptons), which are believed to be the fundamental components of matter, or scalar fields, like the Higgs particles, which are used to introduced mass in a gauge theory (and that, however, could be composed of more fundamental fermion fields).

In the late 19th century with the discovery of the electron, and in the early 20th century, with the discovery of the atomic nucleus, and the birth of particle physics, matter was seen as made up of electrons, protons and neutrons interacting to form atoms. Today, we know that even protons and neutrons are not indivisible, they can be divided into quarks, while electrons are part of a particle family called leptons. Both quarks and leptons are elementary particles, and are currently seen as being the fundamental constituents of matter.

These quarks and leptons interact through four fundamental forces: gravity, electromagnetism, weak interactions, and strong interactions. The Standard Model of particle physics is currently the best explanation for all of physics, but despite decades of efforts, gravity cannot yet be accounted for at the quantum level; it is only described by classical physics (see quantum gravity and graviton). Interactions between quarks and leptons are the result of an exchange of force-carrying particles (such as photons) between quarks and leptons. The force-carrying particles are not themselves building blocks. As one consequence, mass and energy (which cannot be created or destroyed) cannot always be related to matter (which can be created out of non-matter particles such as photons, or even out of pure energy, such as kinetic energy). Force carriers are usually not considered matter: the carriers of the electric force (photons) possess energy (see Planck relation) and the carriers of the weak force (W and Z bosons) are massive, but neither are considered matter either. However, while these particles are not considered matter, they do contribute to the total mass of atoms, subatomic particles, and all systems that contain them.

SUMMARY

The modern conception of matter has been refined many times in history, in light of the improvement in knowledge of just what the basic building blocks are, and in how they interact. The term "matter" is used throughout physics in a bewildering variety of contexts: for example, one refers to "condensed matter physics", "elementary matter", "partonic" matter, "dark" matter, "anti"-matter, "strange" matter, and "nuclear" matter. In discussions of matter and antimatter, normal matter has been referred to by Alfvén as koinomatter (Gk. common matter). It is fair to say that in physics, there is no broad consensus as to a general definition of matter, and the term "matter" usually is used in conjunction with a specifying modifier.

The history of the concept of matter is a history of the fundamental length scales used to define matter. Different building blocks apply depending upon whether one defines matter on an atomic or elementary particle level. One may use a definition that matter is atoms, or that matter is hadrons, or that matter is leptons and quarks depending upon the scale at which one wishes to define matter.

These quarks and leptons interact through four fundamental forces: gravity, electromagnetism, weak interactions, and strong interactions. The Standard Model of particle physics is currently the best explanation for all of physics, but despite decades of efforts, gravity cannot yet be accounted for at the quantum level; it is only described by classical physics (see quantum gravity and graviton).

REFERENCES

1. R. Penrose (1991). "The mass of the classical vacuum". In S. Saunders, H.R. Brown. The Philosophy of Vacuum. Oxford University Press. p. 21. ISBN 0-19-824449-5.
2. "Matter (physics)". McGraw-Hill's Access Science: Encyclopedia of Science and Technology Online. Retrieved 2009-05-24.
3. P. Davies (1992). The New Physics: A Synthesis. Cambridge University Press. p. 1. ISBN 0-521-43831-4.

4. G. ‘t Hooft (1997). In search of the ultimate building blocks. Cambridge University Press. p. 6. ISBN 0-521-57883-3.
5. “RHIC Scientists Serve Up “Perfect” Liquid” (Press release). Brookhaven National Laboratory. 18 April 2005. Retrieved 2009-09-15.
6. J. Olmsted; G.M. Williams (1996). Chemistry: The Molecular Science (2nd ed.). Jones & Bartlett. p. 40. ISBN 0-8151-8450-6.
7. J. Mongillo (2007). Nanotechnology 101. Greenwood Publishing. p. 30. ISBN 0-313-33880-9.
8. P.C.W. Davies (1979). The Forces of Nature. Cambridge University Press. p. 116. ISBN 0-521-22523-X.
9. S. Weinberg (1998). The Quantum Theory of Fields. Cambridge University Press. p. 2. ISBN 0-521-55002-5.
10. M. Masujima (2008). Path Integral Quantization and Stochastic Quantization. Springer. p. 103. ISBN 3-540-87850-5.
11. S.M. Walker; A. King (2005). What is Matter?. Lerner Publications. p. 7. ISBN 0-8225-5131-4.
12. J.Kenkel; P.B. Kelter; D.S. Hage (2000). Chemistry: An Industry-based Introduction with CD-ROM. CRC Press. p. 2. ISBN 1-56670-303-4. All basic science textbooks define matter as simply the collective aggregate of all material substances that occupy space and have mass or weight.
13. K.A. Peacock (2008). The Quantum Revolution: A Historical Perspective. Greenwood Publishing Group. p. 47. ISBN 0-313-33448-X.
14. M.H. Krieger (1998). Constitutions of Matter: Mathematically Modeling the Most Everyday of Physical Phenomena. University of Chicago Press. p. 22. ISBN 0-226-45305-7.
15. G. F. Barker (1870). “Divisions of matter”. A text-book of elementary chemistry: theoretical and inorganic. John F Morton & Co. p. 2. ISBN 978-1-4460-2206-1.
16. M. de Podesta (2002). Understanding the Properties of Matter (2nd ed.). CRC Press. p. 8. ISBN 0-415-25788-3.
17. B. Povh; K. Rith; C. Scholz; F. Zetsche; M. Lavelle (2004). “Part I: Analysis: The building blocks of matter”. Particles and Nuclei: An Introduction to the Physical Concepts (4th ed.). Springer. ISBN 3-540-20168-8.
18. B. Carithers; P. Grannis (1995). “Discovery of the Top Quark”(PDF). Beam Line. SLAC National Accelerator Laboratory. 25 (3): 4–16.
19. D. Green (2005). High PT physics at hadron colliders. Cambridge University Press. p. 23. ISBN 0-521-83509-7.
20. L. Smolin (2007). The Trouble with Physics: The Rise of String Theory, the Fall of a Science, and What Comes Next. Mariner Books. p. 67. ISBN 0-618-91868-X.
21. The W boson mass is 80.398 GeV; see Figure 1 in C. Amsler et al. (Particle Data Group) (2008). “Review of Particle Physics: The Mass and Width of the

W Boson" (PDF). Physics Letters B. 667: 1. Bibcode:2008PhLB..667....1A. doi:10.1016/j.physletb.2008.07.018.

22. I.J.R. Aitchison; A.J.G. Hey (2004). Gauge Theories in Particle Physics. CRC Press. p. 48. ISBN 0-7503-0864-8.
23. B. Povh; K. Rith; C. Scholz; F. Zetsche; M. Lavelle (2004). Particles and Nuclei: An Introduction to the Physical Concepts. Springer. p. 103. ISBN 3-540-20168-8.
24. A.M. Green (2004). Hadronic Physics from Lattice QCD. World Scientific. p. 120. ISBN 981-256-022-X.
25. T. Hatsuda (2008). "Quark–gluon plasma and QCD". In H. Akai. Condensed matter theories. 21. Nova Publishers. p. 296. ISBN 1-60021-501-7.
26. K.W Staley (2004). "Origins of the Third Generation of Matter". The Evidence for the Top Quark. Cambridge University Press. p. 8. ISBN 0-521-82710-8.
27. Y. Ne'eman; Y. Kirsh (1996). The Particle Hunters (2nd ed.). Cambridge University Press. p. 276. ISBN 0-521-47686-0. [T]he most natural explanation to the existence of higher generations of quarks and leptons is that they correspond to excited states of the first generation, and experience suggests that excited systems must be composite
28. S.M. Caroll (2004). Spacetime and Geometry. Addison Wesley. pp. 163–164. ISBN 0-8053-8732-3.
29. P. Davies (1992). The New Physics: A Synthesis. Cambridge University Press. p. 499. ISBN 0-521-43831-4. Matter fields: the fields whose quanta describe the elementary particles that make up the material content of the Universe (as opposed to the gravitons and their supersymmetric partners).
30. C. Amsler et al. (Particle Data Group) (2008). "Reviews of Particle Physics: Quarks" (PDF). Physics Letters B. 667: 1. Bibcode:2008PhLB..667....1A. doi:10.1016/j.physletb.2008.07.018.
31. "Five Year Results on the Oldest Light in the Universe". NASA. 2008. Retrieved 2008-05-02.
32. Persic, Massimo; Salucci, Paolo (1992-09-01). "The baryon content of the Universe". Monthly Notices of the Royal Astronomical Society. 258 (1): 14P–18P. arXiv:astro-ph/0502178 . Bibcode:1992MNRAS.258P..14P. doi:10.1093/mnras/258.1.14P. ISSN 0035-8711.
33. H.S. Goldberg; M.D. Scadron (1987). Physics of Stellar Evolution and Cosmology. Taylor & Francis. p. 202. ISBN 0-677-05540-4.
34. H.S. Goldberg; M.D. Scadron (1987). Physics of Stellar Evolution and Cosmology. Taylor & Francis. p. 233. ISBN 0-677-05540-4.
35. J.-P. Luminet; A. Bullough; A. King (1992). Black Holes. Cambridge University Press. p. 75. ISBN 0-521-40906-3.
36. A. Bodmer (1971). "Collapsed Nuclei". Physical Review D. 4 (6): 1601. Bibcode:1971PhRvD...4.1601B. doi:10.1103/PhysRevD.4.1601.
37. E. Witten (1984). "Cosmic Separation of Phases". Physical Review D. 30 (2): 272. Bibcode:1984PhRvD..30..272W. doi:10.1103/PhysRevD.30.272.

38. C. Amsler et al. (Particle Data Group) (2008). "Review of Particle Physics: Leptons" (PDF). Physics Letters B. 667: 1. Bibcode:2008PhLB..667....1A. doi:10.1016/j.physletb.2008.07.018.
39. C. Amsler et al. (Particle Data Group) (2008). "Review of Particle Physics: Neutrinos Properties" (PDF). Physics Letters B. 667: 1. Bibcode:2008PhLB..667....1A. doi:10.1016/j.physletb.2008.07.018.
40. S. R. Logan (1998). Physical Chemistry for the Biomedical Sciences. CRC Press. pp. 110–111. ISBN 0-7484-0710-3.
41. P.J. Collings (2002). "Chapter 1: States of Matter". Liquid Crystals: Nature's Delicate Phase of Matter. Princeton University Press. ISBN 0-691-08672-9.
42. D.H. Trevena (1975). "Chapter 1.2: Changes of phase". The Liquid Phase. Taylor & Francis. ISBN 978-0-85109-031-3.
43. National Research Council (US) (2006). Revealing the hidden nature of space and time. National Academies Press. p. 46. ISBN 0-309-10194-8.
44. J.P. Ostriker; P.J. Steinhardt (2003). "New Light on Dark Matter". Science. 300 (5627): 1909–13. arXiv:astro-ph/0306402 . Bibcode:2003Sci...300.1909O. doi:10.1126/science.1085976. PMID 12817140.
45. K. Pretzl (2004). "Dark Matter, Massive Neutrinos and Susy Particles". Structure and Dynamics of Elementary Matter. Walter Greiner. p. 289. ISBN 1-4020-2446-0.
46. K. Freeman; G. McNamara (2006). "What can the matter be?". In Search of Dark Matter. Birkhäuser Verlag. p. 105. ISBN 0-387-27616-5.
47. J.C. Wheeler (2007). Cosmic Catastrophes: Exploding Stars, Black Holes, and Mapping the Universe. Cambridge University Press. p. 282. ISBN 0-521-85714-7.
48. J. Gribbin (2007). The Origins of the Future: Ten Questions for the Next Ten Years. Yale University Press. p. 151. ISBN 0-300-12596-8.
49. P. Schneider (2006). Extragalactic Astronomy and Cosmology. Springer. p. 4, Fig. 1.4. ISBN 3-540-33174-3.
50. T. Koupelis; K.F. Kuhn (2007). In Quest of the Universe. Jones & Bartlett Publishers. p. 492; Fig. 16.13. ISBN 0-7637-4387-9.
51. M. H. Jones; R. J. Lambourne; D. J. Adams (2004). An Introduction to Galaxies and Cosmology. Cambridge University Press. p. 21; Fig. 1.13. ISBN 0-521-54623-0.
52. D. Majumdar (2007). "Dark matter — possible candidates and direct detection". arXiv:hep-ph/0703310 [hep-ph].
53. K.A. Olive (2003). "Theoretical Advanced Study Institute lectures on dark matter". arXiv:astro-ph/0301505 [astro-ph].
54. K.A. Olive (2009). "Colliders and Cosmology". European Physical Journal C. 59 (2): 269–295. arXiv:0806.1208 . Bibcode:2009EPJC...59..269O. doi:10.1140/epjc/s10052-008-0738-8.
55. J.C. Wheeler (2007). Cosmic Catastrophes. Cambridge University Press. p. 282. ISBN 0-521-85714-7.

56. L. Smolin (2007). The Trouble with Physics. Mariner Books. p. 16. ISBN 0-618-91868-X.

57. S. Toulmin; J. Goodfield (1962). The Architecture of Matter. University of Chicago Press. pp. 48–54.

58. Discussed by Aristotle in Physics, esp. book I, but also later; as well as Metaphysics I–II.

59. For a good explanation and elaboration, see R.J. Connell (1966). Matter and Becoming. Priory Press.

60. H. G. Liddell; R. Scott; J. M. Whiton (1891). A lexicon abridged from Liddell & Scott's Greek–English lexicon. Harper and Brothers. p. 72.

61. R. Descartes (1644). "The Principles of Human Knowledge". Principles of Philosophy I. p. 53.

62. though even this property seems to be non-essential (René Descartes, Principles of Philosophy II [1644], "On the Principles of Material Things", no. 4.)

63. R. Descartes (1644). "The Principles of Human Knowledge". Principles of Philosophy I. pp. 8, 54, 63.

64. D.L. Schindler (1986). "The Problem of Mechanism". In D.L. Schindler. Beyond Mechanism. University Press of America.

65. E.A. Burtt, Metaphysical Foundations of Modern Science (Garden City, New York: Doubleday and Company, 1954), 117–118.

66. J.E. McGuire and P.M. Heimann, "The Rejection of Newton's Concept of Matter in the Eighteenth Century", The Concept of Matter in Modern Philosophy ed. Ernan McMullin (Notre Dame: University of Notre Dame Press, 1978), 104–118 (105).

67. Isaac Newton, Mathematical Principles of Natural Philosophy, trans. A. Motte, revised by F. Cajori (Berkeley: University of California Press, 1934), pp. 398–400. Further analyzed by Maurice A. Finocchiaro, "Newton's Third Rule of Philosophizing: A Role for Logic in Historiography", Isis 65:1 (Mar. 1974), pp. 66–73.

68. Isaac Newton, Optics, Book III, pt. 1, query 31.

69. McGuire and Heimann, 104.

70. N. Chomsky (1988). Language and problems of knowledge: the Managua lectures (2nd ed.). MIT Press. p. 144. ISBN 0-262-53070-8.

71. McGuire and Heimann, 113.

72. Nevertheless, it remains true that the mathematization regarded as requisite for a modern physical theory carries its own implicit notion of matter, which is very like Descartes', despite the demonstrated vacuity of the latter's notions.

73. M. Wenham (2005). Understanding Primary Science: Ideas, Concepts and Explanations (2nd ed.). Paul Chapman Educational Publishing. p. 115. ISBN 1-4129-0163-4.

74. J.C. Maxwell (1876). Matter and Motion. Society for Promoting Christian Knowledge. p. 18. ISBN 0-486-66895-9.

75. T.H. Levere (1993). "Introduction". Affinity and Matter: Elements of Chemical Philosophy, 1800–1865. Taylor & Francis. ISBN 2-88124-583-8.

76. G.F. Barker (1870). "Introduction". A Text Book of Elementary Chemistry: Theoretical and Inorganic. John P. Morton and Company. p. 2.

77. J. J. Thomson (1909). "Preface". Electricity and Matter. A. Constable.

78. O.W. Richardson (1914). "Chapter 1". The Electron Theory of Matter. The University Press.

79. M. Jacob (1992). The Quark Structure of Matter. World Scientific. ISBN 981-02-3687-5.

80. V. de Sabbata; M. Gasperini (1985). Introduction to Gravitation. World Scientific. p. 293. ISBN 9971-5-0049-3.

81. The history of the concept of matter is a history of the fundamental length scales used to define matter. Different building blocks apply depending upon whether one defines matter on an atomic or elementary particle level. One may use a definition that matter is atoms, or that matter is hadrons, or that matter is leptons and quarks depending upon the scale at which one wishes to define matter. B. Povh; K. Rith; C. Scholz; F. Zetsche; M. Lavelle (2004). "Fundamental constituents of matter". Particles and Nuclei: An Introduction to the Physical Concepts (4th ed.). Springer. ISBN 3-540-20168-8.

82. Allday (2001). Quarks, Leptons and the Big Bang. CRC Press. p. 12. ISBN 0-7503-0806-0.

83. B.A. Schumm (2004). Deep Down Things: The Breathtaking Beauty of Particle Physics. Johns Hopkins University Press. p. 57. ISBN 0-8018-7971-X.

84. M. Jibu; K. Yasue (1995). Quantum Brain Dynamics and Consciousness. John Benjamins Publishing Company. p. 62. ISBN 1-55619-183-9., B. Martin (2009). Nuclear and Particle Physics(2nd ed.). John Wiley & Sons. p. 125. ISBN 0-470-74275-5. and K. W. Plaxco; M. Gross (2006). Astrobiology: A Brief Introduction. Johns Hopkins University Press. p. 23. ISBN 0-8018-8367-9.

85. P. A. Tipler; R. A. Llewellyn (2002). Modern Physics. Macmillan. pp. 89–91, 94–95. ISBN 0-7167-4345-0.

86. P. Schmüser; H. Spitzer (2002). "Particles". In L. Bergmann; et al. Constituents of Matter: Atoms, Molecules, Nuclei. CRC Press. pp. 773 ff. ISBN 0-8493-1202-7.

87. P. M. Chaikin; T. C. Lubensky (2000). Principles of Condensed Matter Physics. Cambridge University Press. p. xvii. ISBN 0-521-79450-1.

88. W. Greiner (2003). W. Greiner; M.G. Itkis; G. Reinhardt; M.C. Güçlü, eds. Structure and Dynamics of Elementary Matter. Springer. p. xii. ISBN 1-4020-2445-2.

89. P. Sukys (1999). Lifting the Scientific Veil: Science Appreciation for the Nonscientist. Rowman & Littlefield. p. 87. ISBN 0-8476-9600-6.

90. B. Povh; K. Rith; C. Scholz; F. Zetsche; M. Lavelle (2004). "Fundamental constituents of matter". Particles and Nuclei: An Introduction to the Physical Concepts (4th ed.). Springer. ISBN 3-540-20168-8.

CHAPTER 4

Chemical Compound

Chapter Contents

- **Introduction**
- **Chemical Element**
- **Chemical Revolution**
- **Chemical Structure**
- **IUPAC Nomenclature Of Inorganic Chemistry**

A chemical compound is a type of molecule (collection of bonded atoms) made up of more than one element. For instance, water (H_2O) is a compound, whereas diatomic hydrogen (H_2) is not. Most of the atoms that make up the Earth's crust and living organisms are in compound form. Compounds are represented by molecular formulae which list the constituent atoms and their abundance.

INTRODUCTION

A chemical compound (or just compound if used in the context of chemistry) is an entity consisting of two or more atoms, at least two from different elements, which associate via chemical bonds. There are four types of compounds, depending on how the constituent atoms are held together: molecules held together by covalent bonds, salts held together by ionic bonds, intermetallic compounds held together by metallic bonds, and certain complexes held together by coordinate covalent bonds. Many chemical compounds have a unique numerical identifier assigned by the Chemical Abstracts Service (CAS): its CAS number.

A chemical formula is a way of expressing information about the proportions of atoms that constitute a particular chemical compound, using the standard abbreviations for the chemical elements, and subscripts to indicate the number of atoms involved. For example, water is composed of two hydrogen atoms bonded to one oxygen atom: the chemical formula is H_2O.

A compound can be converted to a different chemical composition by interaction with a second chemical compound via a chemical reaction. In this process, bonds between atoms are broken in both of the interacting compounds, and then bonds are reformed so that new associations are made between atoms. Schematically, this reaction could be described as AB + CD --> AC + BD, where A, B, C, and D are each unique atoms; and AB, CD, AC, and BD are each unique compounds.

A chemical element bonded to an identical chemical element is not a chemical compound since only one element, not two different elements, is involved. Examples are the diatomic molecule hydrogen (H_2) and the polyatomic molecule sulfur (S_8).

Definitions

Any substance consisting of two or more different types of atoms (chemical elements) in a fixed proportion of its atoms (i.e., stoichiometry) can be termed a chemical compound; the concept is most readily understood when considering pure chemical substances.:15 It follows from their being composed of fixed proportions of two or more types of atoms that chemical compounds can be converted, via chemical reaction, into compounds or substances each having fewer atoms. In the case of non-stoichiometric compounds, the proportions may be reproducible with regard to their preparation, and give fixed proportions of their component elements, but proportions that are not integral. Chemical compounds have a unique and defined chemical structure held together in a defined spatial arrangement by chemical bonds. Chemical compounds can be molecular compounds held together by covalent bonds, salts held together by ionic bonds, intermetallic compounds held together by metallic bonds, or the subset of chemical complexes that are held together by coordinate covalent bonds. Pure chemical elements are generally not considered chemical compounds, failing the two or more atom requirement, though they often consist of molecules composed of multiple atoms (such as in the diatomic molecule H_2, or the polyatomic molecule S_8, etc.).

There is varying and sometimes inconsistent nomenclature differentiating substances, which include truly non-stoichiometric examples, from chemical compounds, which require the fixed ratios. Many solid chemical substances—for example many silicate minerals—are chemical substances, but do not have simple formulae reflecting chemically bonding of elements to one another in fixed ratios; even so, these crystalline substances are often called "non-stoichiometric compounds". It may be argued that they are related to, rather than being chemical compounds, insofar as the variability in their compositions is often due to either the presence of foreign elements trapped within the crystal structure of an otherwise known true chemical compound, or due to perturbations in structure relative to the known compound that arise because of an excess of deficit of the constituent elements at places in its structure; such non-stoichiometric substances form most of the crust and mantle of the Earth. Other compounds regarded as chemically identical may have varying amounts of heavy or light isotopes of the constituent elements, which changes the ratio of elements by mass slightly.

Elementary Concepts

Characteristic properties of compounds include that elements in a compound are present in a definite proportion. For example, the molecule of the compound water is composed of hydrogen and oxygen in a ratio of 2:1. In addition, compounds have a definite set of properties, and the elements that comprise a compound do not retain their original properties. For example, hydrogen, which is combustible and non-supportive of combustion, combines with oxygen, which is non-combustible and supportive of combustion, to produce the compound water, which is non-combustible and non-supportive of combustion.

Comparison to Mixtures

The physical and chemical properties of compounds differ from those of their constituent elements. This is one of the main criteria that distinguish a compound from a mixture of elements or other substances—in general, a mixture's properties are closely related to, and depend on, the properties of its constituents. Another criterion that distinguishes a compound from a mixture is that constituents of a mixture can usually be separated by simple mechanical means, such as filtering, evaporation, or magnetic force, but components of a compound can be separated only by a chemical reaction. However, mixtures can be created by mechanical means alone, but a compound can be created (either from elements or from other compounds, or a combination of the two) only by a chemical reaction.

Some mixtures are so intimately combined that they have some properties similar to compounds and may easily be mistaken for compounds. One example is alloys. Alloys are made mechanically, most commonly by heating the constituent metals to a liquid state, mixing them thoroughly, and then cooling the mixture quickly so that the constituents are trapped in the base metal. Other examples of compound-like mixtures include intermetallic compounds and solutions of alkali metals in a liquid form of ammonia.

Formula

A chemical formula is a way of expressing information about the proportions of atoms that constitute a particular chemical compound, using a single line of chemical element symbols, numbers, and sometimes also other symbols, such as parentheses, dashes, brackets, commas and plus (+) and minus (−) signs.

Compounds may be described using formulas in various formats. For compounds that exist as molecules, the formula for the molecular unit is shown. For polymeric materials, such as minerals and many metal oxides, the empirical formula is normally given, e.g. NaCl for table salt.

The elements in a chemical formula are normally listed in a specific order, called the Hill system. In this system, the carbon atoms (if there are any) are usually listed first, any hydrogen atoms are listed next, and all other elements follow in alphabetical order. If the formula contains no carbon, then all of the elements, including hydrogen, are listed alphabetically. There are, however, several important exceptions to the normal rules. For ionic compounds, the positive ion is almost always listed first and the negative ion is listed second. For oxides, oxygen is usually listed last.

In general, organic acids follow the normal rules with C and H coming first in the formula. For example, the formula for trifluoroacetic acid is usually written as $C_2HF_3O_2$. More descriptive formulas can convey structural information, such as writing the formula for trifluoroacetic acid as CF_3CO_2H. On the other hand, the chemical formulas for most inorganic acids and bases are exceptions to the normal rules. They are written according to the rules for ionic compounds (positive first, negative second), but they also follow rules that emphasize their Arrhenius definitions. To be specific, the formula for most inorganic acids begins with hydrogen and the formula for most bases ends with the hydroxide ion (OH^-). Formulas for inorganic compounds do not often convey structural information, as illustrated by the common use of the formula H_2SO_4 for a molecule (sulfuric acid) that contains no H-S bonds. A more descriptive presentation would be $O_2S(OH)_2$, but it is almost never written this way.

Phases and Thermal Properties

Compounds may have several possible phases. All compounds can exist as solids, at least at low enough temperatures. Molecular compounds may also exist as liquids, gases, and, in some cases, even plasmas. All compounds decompose upon applying heat. The temperature at which such fragmentation occurs is often called the decomposition temperature. Decomposition temperatures are not sharp and depend on pressure, temperature, and the concentration of each species in the compound.

CHEMICAL ELEMENT

A chemical element or element is a species of atoms having the same number of protons in their atomic nuclei (i.e. the same atomic number, Z). There are 118 elements that have been identified, of which the first 94 occur naturally on Earth with the remaining 24 being synthetic elements. There are 80 elements that have at least one stable isotope and 38 that have exclusively radioactive isotopes, which decay over time into other elements. Iron is the most abundant element (by mass) making up Earth, while oxygen is the most common element in the crust of Earth.

Chemical elements constitute all of the ordinary matter of the universe. However astronomical observations suggest that ordinary observable matter is only approximately 15% of the matter in the universe: the remainder is dark matter, the composition of which is unknown, but it is not composed of chemical elements. The two lightest elements, hydrogen and helium were mostly formed in the Big Bang and are the most common elements in the universe. The next three elements (lithium, beryllium and boron) were formed mostly by cosmic ray spallation, and are thus more rare than those that follow. Formation of elements with from six to twenty six protons occurred and continues to occur in main sequence stars via stellar nucleosynthesis. The high abundance of oxygen, silicon, and iron on Earth reflects their common production in such stars. Elements with greater than twenty-six protons are formed by supernova nucleosynthesis in supernovae, which, when they explode, blast these elements far into space as supernova remnants, where they may become incorporated into planets when they are formed.

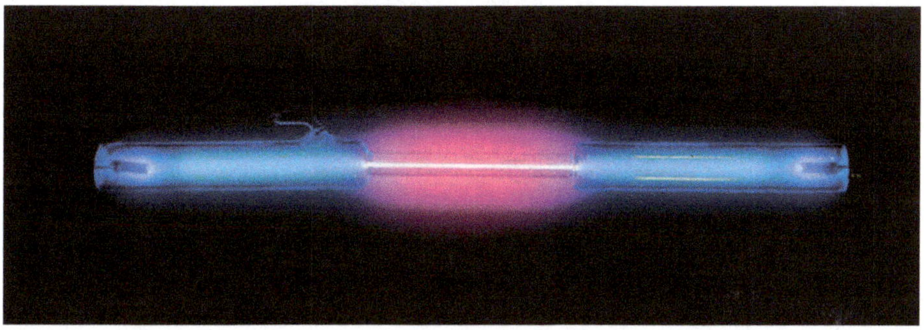

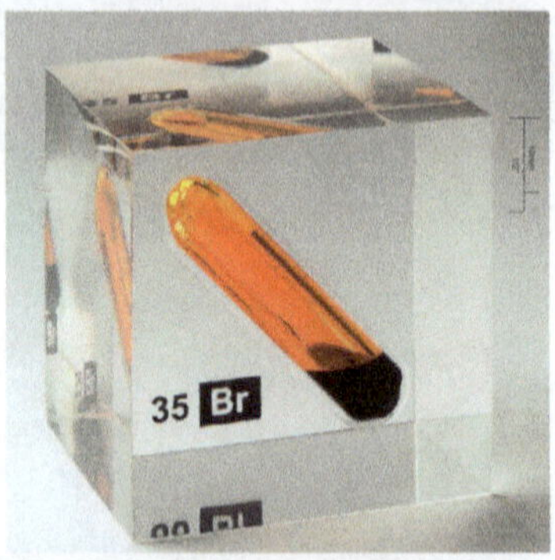

Top: The periodic table of the chemical elements. Below: Examples of certain chemical elements. From left to right: hydrogen, barium, copper, uranium, bromine, and helium.

The term "element" is used for atoms with a given number of protons (regardless of whether or not they are ionized or chemically bonded, e.g. hydrogen in water) as well as for a pure chemical substance consisting of a single element (e.g. hydrogen gas). For the second meaning, the terms "elementary substance" and "simple substance" have been suggested, but they have not gained much acceptance in the English-language chemical literature, whereas in some other languages their equivalent is widely used (e.g. French corps simple, Russian простое вещество). One element can form multiple substances differing in their structure; they are called allotropes of the element.

When different elements are chemically combined, with the atoms held together by chemical bonds, they form chemical compounds. Only a minority of elements are found uncombined as relatively pure minerals. Among the more common of such "native elements" are copper, silver, gold, carbon (as coal, graphite, or diamonds), and sulfur. All but a few of the most inert elements, such as noble gases and noble metals, are usually found on Earth in chemically combined form, as chemical compounds. While about 32 of the chemical elements occur on Earth in native uncombined forms, most of these occur as mixtures. For example, atmospheric air is primarily a mixture of nitrogen, oxygen, and argon, and native solid elements occur in alloys, such as that of iron and nickel.

The history of the discovery and use of the elements began with primitive human societies that found native elements like carbon, sulfur, copper and gold. Later civilizations extracted elemental copper, tin, lead and iron from their ores by smelting, using charcoal. Alchemists and chemists subsequently identified many more, with almost all of the naturally-occurring elements becoming known by 1900.

The properties of the chemical elements are summarized on the periodic table, which organizes the elements by increasing atomic number into rows ("periods") in which the columns ("groups") share recurring ("periodic") physical and chemical properties. Save for unstable radioactive elements with short half-lives, all of the elements are available industrially, most of them in high degrees of purity.

Description

The lightest chemical elements are hydrogen and helium, both created by Big Bang nucleosynthesis during the first 20 minutes of the universe in a ratio of around 3:1 by mass (or 12:1 by number of atoms), along with tiny traces of the next two elements, lithium and beryllium. Almost all other elements found in nature were made by various natural methods of nucleosynthesis. On Earth, small amounts of new atoms are naturally produced in nucleogenic reactions, or in cosmogenic processes, such as cosmic ray spallation. New atoms are also naturally produced on Earth as radiogenic daughter isotopes of ongoing radioactive decay processes such as alpha decay, beta decay, spontaneous fission, cluster decay, and other rarer modes of decay.

Of the 94 naturally occurring elements, those with atomic numbers 1 through 82 each have at least one stable isotope (except for technetium, element 43 and promethium, element 61, which have no stable isotopes). Isotopes considered stable are those for which no radioactive decay has yet been observed. Elements with atomic numbers 83 through 94 are unstable to the point that radioactive decay of all isotopes can be detected. Some of these elements, notably bismuth (atomic number 83), thorium (atomic number 90), and uranium (atomic number 92), have one or more isotopes with half-lives long enough to survive as remnants of the explosive stellar nucleosynthesis that produced the heavy elements before the formation of our solar system. At over 1.9×10^{19} years, over a billion times longer than the current estimated age of the universe, bismuth-209 (atomic number 83) has the longest known alpha decay half-life of any naturally occurring element, and is almost always considered on par with the 80 stable elements. The very heaviest elements (those beyond plutonium, element 94) undergo radioactive decay with half-lives so short that they are not found in nature and must be synthesized.

As of 2010, there are 118 known elements (in this context, "known" means observed well enough, even from just a few decay products, to have been differentiated from other elements). Of these 118 elements, 94 occur naturally on Earth. Six of these occur in extreme trace quantities: technetium, atomic number 43; promethium, number 61; astatine, number 85; francium, number 87; neptunium, number 93; and plutonium, number 94. These 94 elements have been detected in the universe at large, in the spectra of stars and also supernovae, where short-lived radioactive elements are newly being made. The first 94 elements have been detected directly on Earth as primordial nuclides present from the formation of the solar system, or as naturally-occurring fission or transmutation products of uranium and thorium.

The remaining 24 heavier elements, not found today either on Earth or in astronomical spectra, have been produced artificially: these are all radioactive, with very short half-lives; if any atoms of these elements were present at the formation of Earth, they are extremely likely, to the point of certainty, to have already decayed, and if present in novae, have been in quantities too small to have been noted. Technetium was the first purportedly non-naturally occurring element synthesized, in 1937, although trace amounts of technetium have since been found in nature (and also the element may have been discovered naturally in 1925). This pattern of artificial production and later natural discovery has been repeated with several other radioactive naturally-occurring rare elements.

Lists of the elements are available by name, by symbol, by atomic number, by density, by melting point, and by boiling point as well as ionization energies of the elements. The nuclides of stable and radioactive elements are also available as a list of nuclides, sorted by length of half-life for those that are unstable. One of the most convenient, and certainly the most traditional presentation of the elements, is in the form of the periodic table, which groups together elements with similar chemical properties (and usually also similar electronic structures).

Atomic Number

The atomic number of an element is equal to the number of protons in each atom, and defines the element. For example, all carbon atoms contain 6 protons in their atomic nucleus; so the atomic number of carbon is 6. Carbon atoms may have different numbers of neutrons; atoms of the same element having different numbers of neutrons are known as isotopes of the element.

The number of protons in the atomic nucleus also determines its electric charge, which in turn determines the number of electrons of the atom in its non-ionized state. The electrons are placed into atomic orbitals that determine the atom's various chemical properties. The number of neutrons in a nucleus usually has very little effect on an element's chemical properties (except in the case of hydrogen and deuterium). Thus, all carbon isotopes have nearly identical chemical properties because they all have six protons and six electrons, even though carbon atoms may, for example, have 6 or 8 neutrons. That is why the atomic number, rather than mass number or atomic weight, is considered the identifying characteristic of a chemical element.

Isotopes

Isotopes are atoms of the same element (that is, with the same number of protons in their atomic nucleus), but having different numbers of neutrons. Most (66 of 94) naturally occurring elements have more than one stable isotope. Thus, for example, there are three main isotopes of carbon. All carbon atoms have 6 protons in the nucleus, but they can have either 6, 7, or 8 neutrons. Since the mass numbers of these are 12, 13 and 14 respectively, the three isotopes of carbon are known as carbon-12, carbon-13, and carbon-14, often abbreviated to ^{12}C, ^{13}C, and ^{14}C. Carbon in everyday life and in chemistry is a mixture of ^{12}C (about 98.9%), ^{13}C (about 1.1%) and about 1 atom per trillion of ^{14}C.

Except in the case of the isotopes of hydrogen (which differ greatly from each other in relative mass—enough to cause chemical effects), the isotopes of a given element are chemically nearly indistinguishable.

All of the elements have some isotopes that are radioactive (radioisotopes), although not all of these radioisotopes occur naturally. The radioisotopes typically decay into other elements upon radiating an alpha or beta particle. If an element has isotopes that are not radioactive, these are termed "stable" isotopes. All of the known stable isotopes occur naturally (see primordial isotope). The many radioisotopes that are not found in nature have been characterized after being artificially made. Certain elements have no stable isotopes and are composed only of radioactive isotopes: specifically the elements without any stable isotopes are technetium (atomic number 43), promethium (atomic number 61), and all observed elements with atomic numbers greater than 82.

Of the 80 elements with at least one stable isotope, 26 have only one single stable isotope. The mean number of stable isotopes for the 80 stable elements is 3.1 stable isotopes per element. The largest number of stable isotopes that occur for a single element is 10 (for tin, element 50).

Isotopic Mass and Atomic Mass

The mass number of an element, A, is the number of nucleons (protons and neutrons) in the atomic nucleus. Different isotopes of a given element are distinguished by their mass numbers, which are conventionally written as a superscript on the left hand side of the atomic symbol (e.g., 238U). The mass number is always a simple whole number and has units of "nucleons." An example of a referral to a mass number is "magnesium-24," which is an atom with 24 nucleons (12 protons and 12 neutrons).

Whereas the mass number simply counts the total number of neutrons and protons and is thus a natural (or whole) number, the atomic mass of a single atom is a real number for the mass of a particular isotope of the element, the unit being u. In general, when expressed in u it differs in value slightly from the mass number for a given nuclide (or isotope) since the mass of the protons and neutrons is not exactly 1 u, since the electrons contribute a lesser share to the atomic mass as neutron number exceeds proton number, and (finally) because of the nuclear binding energy. For example, the atomic mass of chlorine-35 to five significant digits is 34.969 u and that of chlorine-37 is 36.966 u. However, the atomic mass in u of each isotope is quite close to its simple mass number (always within 1%). The only isotope whose atomic mass is exactly a natural number is ^{12}C, which by definition has a mass of exactly 12, because u is defined as 1/12 of the mass of a free neutral carbon-12 atom in the ground state.

The relative atomic mass (historically and commonly also called "atomic weight") of an element is the average of the atomic masses of all the chemical element's isotopes as found in a particular environment, weighted by isotopic abundance, relative to the atomic mass unit (u). This number may be a fraction that is not close to a whole number, due to the averaging process. For example, the relative atomic mass of chlorine is 35.453 u, which differs greatly from a whole number due to being made of an average of 76% chlorine-35 and 24% chlorine-37. Whenever a relative atomic mass value differs by more than 1% from a whole number, it is due to this averaging effect resulting from significant amounts of more than one isotope being naturally present in the sample of the element in question.

Chemically Pure and Isotopically Pure

Chemists and nuclear scientists have different definitions of a pure element. In chemistry, a pure element means a substance whose atoms all (or in practice almost all) have the same atomic number, or number of protons. Nuclear scientists, however, define a pure element as one that consists of only one stable isotope.

For example, a copper wire is 99.99% chemically pure if 99.99% of its atoms are copper, with 29 protons each. However it is not isotopically pure since ordinary copper consists of two stable isotopes, 69% ^{63}Cu and 31% ^{65}Cu, with different numbers of neutrons. However, a pure gold ingot would be both chemically and isotopically pure, since ordinary gold consists only of one isotope, ^{197}Au.

Allotropes

Atoms of chemically pure elements may bond to each other chemically in more than one way, allowing the pure element to exist in multiple structures (spatial arrangements of atoms), known as allotropes, which differ in their properties. For example, carbon can be found as diamond, which has a tetrahedral structure around each carbon atom; graphite, which has layers of carbon atoms with a hexagonal structure stacked on top of each other; graphene, which is a single layer of graphite that is very strong; fullerenes, which have nearly spherical shapes; and carbon nanotubes, which are tubes with a hexagonal structure (even these may differ from each other in electrical properties). The ability of an element to exist in one of many structural forms is known as 'allotropy'.

The standard state, also known as reference state, of an element is defined as its thermodynamically most stable state at 1 bar at a given temperature (typically at 298.15 K). In thermochemistry, an element is defined to have an enthalpy of formation of zero in its standard state. For example, the reference state for carbon is graphite, because the structure of graphite is more stable than that of the other allotropes.

Properties

Several kinds of descriptive categorizations can be applied broadly to the elements, including consideration of their general physical and chemical properties, their states of matter under familiar conditions, their melting and boiling points, their densities, their crystal structures as solids, and their origins.

General Properties

Several terms are commonly used to characterize the general physical and chemical properties of the chemical elements. A first distinction is between metals, which readily conduct electricity, nonmetals, which do not, and a small group, (the metalloids), having intermediate properties and often behaving as semiconductors.

A more refined classification is often shown in colored presentations of the periodic table. This system restricts the terms "metal" and "nonmetal" to only certain of the more broadly defined metals and nonmetals, adding additional terms for certain sets of the more broadly viewed metals and nonmetals. The version of this classification used in the periodic tables presented here includes: actinides, alkali metals, alkaline earth metals, halogens, lanthanides, transition metals, post-transition metals, metalloids, polyatomic nonmetals, diatomic nonmetals, and noble gases. In this system, the alkali metals, alkaline earth metals, and transition metals, as well as the lanthanides and the actinides, are special groups of the metals viewed in a broader sense. Similarly, the polyatomic nonmetals, diatomic nonmetals and the noble gases are nonmetals viewed in the broader sense. In some presentations, the halogens are not distinguished, with astatine identified as a metalloid and the others identified as nonmetals.

States of Matter

Another commonly used basic distinction among the elements is their state of matter (phase), whether solid, liquid, or gas, at a selected standard temperature and pressure (STP). Most of the elements are solids at conventional temperatures and atmospheric pressure, while several are gases. Only bromine and mercury are liquids at 0 degrees Celsius (32 degrees Fahrenheit) and normal atmospheric pressure; caesium and gallium are solids at that temperature, but melt at 28.4 °C (83.2 °F) and 29.8 °C (85.6 °F), respectively.

Melting and Boiling Points

Melting and boiling points, typically expressed in degrees Celsius at a pressure of one atmosphere, are commonly used in characterizing the various elements. While known for most elements, either or both of these measurements is still undetermined for some of the radioactive elements available in only tiny quantities. Since helium remains a liquid even at absolute zero at atmospheric pressure, it has only a boiling point, and not a melting point, in conventional presentations.

Densities

The density at a selected standard temperature and pressure (STP) is frequently used in characterizing the elements. Density is often expressed in grams per cubic centimeter (g/cm3). Since several elements are gases at commonly encountered temperatures, their densities are usually stated for their gaseous forms; when liquefied or solidified, the gaseous elements have densities similar to those of the other elements.

When an element has allotropes with different densities, one representative allotrope is typically selected in summary presentations, while densities for each allotrope can be stated where more detail is provided. For example, the three familiar allotropes of carbon (amorphous carbon, graphite, and diamond) have densities of 1.8–2.1, 2.267, and 3.515 g/cm^3, respectively.

Crystal Structures

The elements studied to date as solid samples have eight kinds of crystal structures: cubic, body-centered cubic, face-centered cubic, hexagonal, monoclinic, orthorhombic, rhombohedral, and tetragonal. For some of the synthetically produced transuranic elements, available samples have been too small to determine crystal structures.

Occurrence and Origin on Earth

Chemical elements may also be categorized by their origin on Earth, with the first 94 considered naturally occurring, while those with atomic numbers beyond 94 have only been produced artificially as the synthetic products of man-made nuclear reactions.

Of the 94 naturally occurring elements, 84 are considered primordial and either stable or weakly radioactive. The remaining 10 naturally occurring elements possess half lives too short for them to have been present at the beginning of the Solar System, and are therefore considered transient elements. (Plutonium is usually also considered a transient element because primordial plutonium has by now decayed to almost undetectable traces.) Of these 10 transient elements, 5 (polonium, radon, radium, actinium, and protactinium) are relatively common decay products of thorium, uranium, and plutonium. The remaining 6 transient elements (technetium, promethium, astatine, francium, neptunium, and plutonium) occur only rarely, as products of rare decay modes or nuclear reaction processes involving uranium or other heavy elements.

Elements with atomic numbers 1 through 40 are all stable, while those with atomic numbers 41 through 82 (except technetium and promethium) are metastable. The half-lives of these metastable "theoretical radionuclides" are so long (at least 100 million times longer than the estimated age of the universe) that their radioactive decay has yet to be detected by experiment. Elements with atomic numbers 83 through 94 are unstable to the point that their radioactive decay can be detected. Three of these elements, bismuth (element 83), thorium (element 90), and uranium (element 92) have one or more isotopes with half-lives long enough to survive as remnants of the explosive stellar nucleosynthesis that produced the heavy elements before the formation of our solar system. For example, at over 1.9×10^{19} years, over a billion times longer than the current estimated age of the universe, bismuth-209 has the longest known alpha decay half-life of any naturally occurring element. The very heaviest 24 elements (those beyond plutonium, element 94) undergo radioactive decay with short half-lives and cannot be produced as daughters of longer-lived elements, and thus they do not occur in nature at all.

The Periodic Table

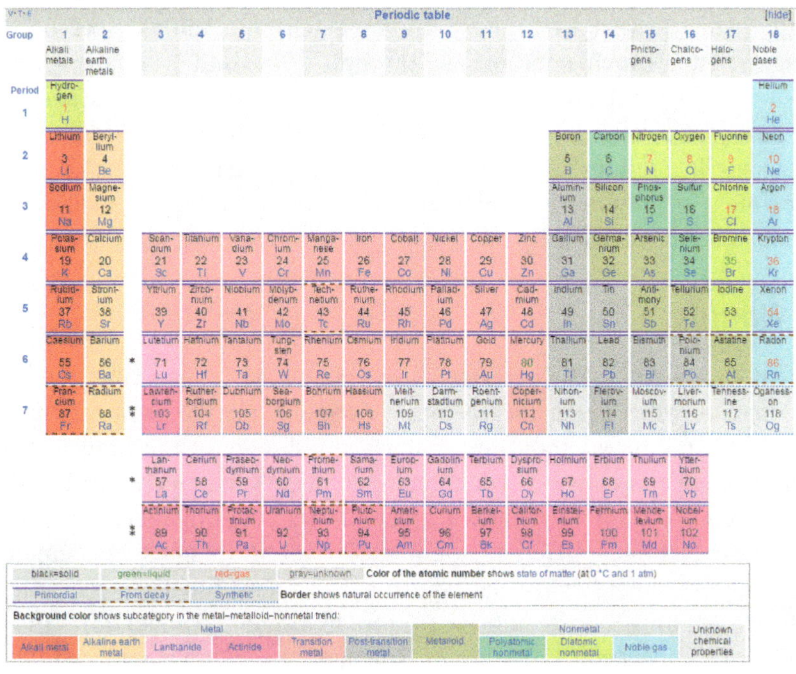

The properties of the chemical elements are often summarized using the periodic table, which powerfully and elegantly organizes the elements by increasing atomic number into rows ("periods") in which the columns ("groups") share recurring ("periodic") physical and chemical properties. The current standard table contains 118 confirmed elements as of 10 April 2010.

Although earlier precursors to this presentation exist, its invention is generally credited to the Russian chemist Dmitri Mendeleev in 1869, who intended the table to illustrate recurring trends in the properties of the elements. The layout of the table has been refined and extended over time as new elements have been discovered and new theoretical models have been developed to explain chemical behavior.

Use of the periodic table is now ubiquitous within the academic discipline of chemistry, providing an extremely useful framework to classify, systematize and compare all the many different forms of chemical behavior. The table has also found wide application in physics, geology, biology, materials science, engineering, agriculture, medicine, nutrition, environmental health, and astronomy. Its principles are especially important in chemical engineering.

Nomenclature and Symbols

The various chemical elements are formally identified by their unique atomic numbers, by their accepted names, and by their symbols.

Atomic Numbers

The known elements have atomic numbers from 1 through 118, conventionally presented as Arabic numerals. Since the elements can be uniquely sequenced by atomic number, conventionally from lowest to highest (as in a periodic table), sets of elements are sometimes specified by such notation as "through", "beyond", or "from ... through", as in "through iron", "beyond uranium", or "from lanthanum through lutetium". The terms "light" and "heavy" are sometimes also used informally to indicate relative atomic numbers (not densities), as in "lighter than carbon" or "heavier than lead", although technically the weight or mass of atoms of an element (their atomic weights or atomic masses) do not always increase monotonically with their atomic numbers.

Element Names

The naming of various substances now known as elements precedes the atomic theory of matter, as names were given locally by various cultures to various minerals, metals, compounds, alloys, mixtures, and other materials, although at the time it was not known which chemicals were elements and which compounds. As they were identified as elements, the existing names for anciently-known elements (e.g., gold, mercury, iron) were kept in most countries. National differences emerged over the names of elements either for convenience, linguistic niceties, or nationalism. For a few illustrative examples: German speakers use "Wasserstoff" (water substance) for "hydrogen", "Sauerstoff" (acid substance) for "oxygen" and "Stickstoff" (smothering substance) for "nitrogen", while English and some romance languages use "sodium" for "natrium" and "potassium" for "kalium", and the French, Italians, Greeks, Portuguese and Poles prefer "azote/azot/azoto" (from roots meaning "no life") for "nitrogen".

For purposes of international communication and trade, the official names of the chemical elements both ancient and more recently recognized are decided by the International Union of Pure and Applied Chemistry (IUPAC), which has decided on a sort of international English language, drawing on traditional English names even when an element's chemical symbol is based on a Latin or other traditional word, for example adopting "gold" rather than "aurum" as the name for the 79th element (Au). IUPAC prefers the British spellings "aluminium" and "caesium" over the U.S. spellings "aluminum" and "cesium", and the U.S. "sulfur" over the British "sulphur". However, elements that are practical to sell in bulk in many countries often still have locally used national names, and countries whose national language does not use the Latin alphabet are likely to use the IUPAC element names.

According to IUPAC, chemical elements are not proper nouns in English; consequently, the full name of an element is not routinely capitalized in English, even if derived from a proper noun, as in californium and einsteinium. Isotope names of chemical elements are also uncapitalized if written out, e.g., carbon-12 or uranium-235. Chemical element symbols (such as Cf for californium and Es for einsteinium), are always capitalized (see below).

In the second half of the twentieth century, physics laboratories became able to produce nuclei of chemical elements with half-lives too short for an appreciable amount of them to exist at any time. These are also named by IUPAC, which generally adopts the name chosen by the discoverer. This practice can lead to the controversial question of which research group actually discovered an element, a question that delayed the naming of elements with atomic number of 104 and higher for a considerable amount of time. (See element naming controversy).

Precursors of such controversies involved the nationalistic namings of elements in the late 19th century. For example, lutetium was named in reference to Paris, France. The Germans were reluctant to relinquish naming rights to the French, often calling it cassiopeium. Similarly, the British discoverer of niobium originally named it columbium, in reference to the New World. It was used extensively as such by American publications prior to the international standardization (in 1950).

Specific Chemical Elements

Before chemistry became a science, alchemists had designed arcane symbols for both metals and common compounds. These were however used as abbreviations in diagrams or procedures; there was no concept of atoms combining to form molecules. With his advances in the atomic theory of matter, John Dalton devised his own simpler symbols, based on circles, to depict molecules.

The current system of chemical notation was invented by Berzelius. In this typographical system, chemical symbols are not mere abbreviations—though each consists of letters of the Latin alphabet. They are intended as universal symbols for people of all languages and alphabets.

The first of these symbols were intended to be fully universal. Since Latin was the common language of science at that time, they were abbreviations based on the Latin names of metals. Cu comes from Cuprum, Fe comes from Ferrum, Ag from Argentum. The symbols were not followed by a period (full stop) as with abbreviations. Later chemical elements were also assigned unique chemical symbols, based on the name of the element, but not necessarily in English. For example, sodium has the chemical symbol 'Na' after the Latin natrium. The same applies to "W" (wolfram) for tungsten, "Fe" (ferrum) for iron, "Hg" (hydrargyrum) for mercury, "Sn" (stannum) for tin, "K" (kalium) for potassium, "Au" (aurum) for gold, "Ag" (argentum) for silver, "Pb" (plumbum) for lead, "Cu" (cuprum) for copper, and "Sb" (stibium) for antimony.

Chemical symbols are understood internationally when element names might require translation. There have sometimes been differences in the past. For example, Germans in the past have used "J" (for the alternate name Jod) for iodine, but now use "I" and "Iod".

The first letter of a chemical symbol is always capitalized, as in the preceding examples, and the subsequent letters, if any, are always lower case (small letters). Thus, the symbols for californium and einsteinium are Cf and Es.

General Chemical Symbols

There are also symbols in chemical equations for groups of chemical elements, for example in comparative formulas. These are often a single capital letter, and the letters are reserved and not used for names of specific elements. For example, an "X" indicates a variable group (usually a halogen) in a class of compounds, while "R" is a radical, meaning a compound structure such as a hydrocarbon chain. The letter "Q" is reserved for "heat" in a chemical reaction. "Y" is also often used as a general chemical symbol, although it is also the symbol of yttrium. "Z" is also frequently used as a general variable group. "E" is used in organic chemistry to denote an electron-withdrawing group or an electrophile; similarly "Nu" denotes a nucleophile. "L" is used to represent a general ligand in inorganic and organometallic chemistry. "M" is also often used in place of a general metal.

At least two additional, two-letter generic chemical symbols are also in informal usage, "Ln" for any lanthanide element and "An" for any actinide element. "Rg" was formerly used for any rare gas element, but the group of rare gases has now been renamed noble gases and the symbol "Rg" has now been assigned to the element roentgenium.

Isotope Symbols

Isotopes are distinguished by the atomic mass number (total protons and neutrons) for a particular isotope of an element, with this number combined with the pertinent element's symbol. IUPAC prefers that isotope symbols be written in superscript notation when practical, for example ^{12}C and ^{235}U. However, other notations, such as carbon-12 and uranium-235, or C-12 and U-235, are also used.

As a special case, the three naturally occurring isotopes of the element hydrogen are often specified as H for ^{1}H (protium), D for ^{2}H (deuterium), and T for ^{3}H (tritium). This convention is easier to use in chemical equations, replacing the need to write out the mass number for each atom. For example, the formula for heavy water may be written D_2O instead of $^{2}H_2O$.

Origin of the Elements

Only about 4% of the total mass of the universe is made of atoms or ions, and thus represented by chemical elements. This fraction is about 15% of the total matter, with the remainder of the matter (85%) being dark matter. The nature of dark matter is unknown, but it is not composed of atoms of chemical elements because it contains no protons, neutrons, or electrons. (The remaining non-matter part of the mass of the universe is composed of the even more mysterious dark energy).

The universe's 94 naturally occurring chemical elements are thought to have been produced by at least four cosmic processes. Most of the hydrogen and helium in the universe was produced primordially in the first few minutes of the Big Bang. Three recurrently occurring later processes are thought to have produced the remaining elements. Stellar nucleosynthesis, an ongoing process, produces all elements from carbon through iron in atomic number, but little lithium, beryllium, or boron. Elements heavier in atomic number than iron, as heavy as uranium and plutonium, are produced by explosive nucleosynthesis in supernovas and other cataclysmic cosmic events. Cosmic ray spallation (fragmentation) of carbon, nitrogen, and oxygen is important to the production of lithium, beryllium and boron.

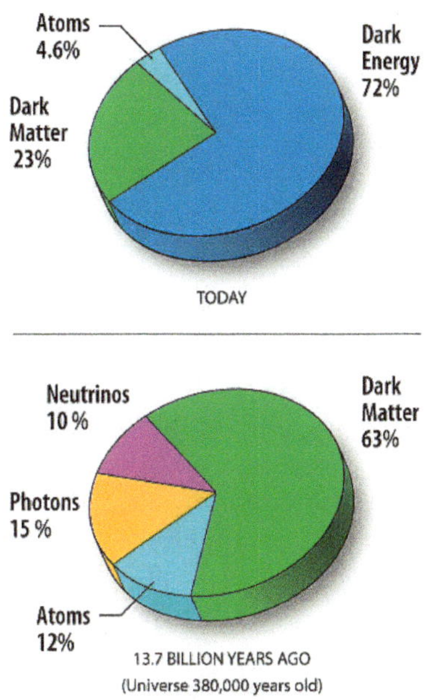

Estimated distribution of dark matter and dark energy in the universe. Only the fraction of the mass and energy in the universe labeled "atoms" is composed of chemical elements.

During the early phases of the Big Bang, nucleosynthesis of hydrogen nuclei resulted in the production of hydrogen^{-1} (protium, 1H) and helium-4 (^{4}He), as well as a smaller amount of deuterium (^{2}H) and very minuscule amounts (on the order of 10^{-10}) of lithium and beryllium. Even smaller amounts of boron may have been produced in the Big Bang, since it has been observed in some very old stars, while carbon has not. It is generally agreed that no heavier elements than boron were produced in the Big Bang. As a result, the primordial abundance of atoms (or ions) consisted of roughly 75% ^{1}H, 25% ^{4}He, and 0.01% deuterium, with only tiny traces of lithium, beryllium, and perhaps boron. Subsequent enrichment of galactic halos occurred due to stellar nucleosynthesis and supernova nucleosynthesis. However, the element abundance in intergalactic space can still closely resemble primordial conditions, unless it has been enriched by some means.

On Earth (and elsewhere), trace amounts of various elements continue to be produced from other elements as products of natural transmutation processes. These include some produced by cosmic rays or other nuclear reactions (see cosmogenic and nucleogenic nuclides), and others produced as decay products of long-lived primordial nuclides. For example, trace (but detectable) amounts of carbon-14 (^{14}C) are continually produced in the atmosphere by cosmic rays impacting nitrogen atoms, and argon-40 (^{40}Ar) is continually produced by the decay of primordially occurring but unstable potassium-40 (^{40}K). Also, three primordially occurring but radioactive actinides, thorium, uranium, and plutonium, decay through a series of recurrently produced but unstable radioactive elements such as radium and radon, which are transiently present in any sample of these metals or their ores or compounds. Three other radioactive elements, technetium, promethium, and neptunium, occur only incidentally in natural materials, produced as individual atoms by natural fission of the nuclei of various heavy elements or in other rare nuclear processes.

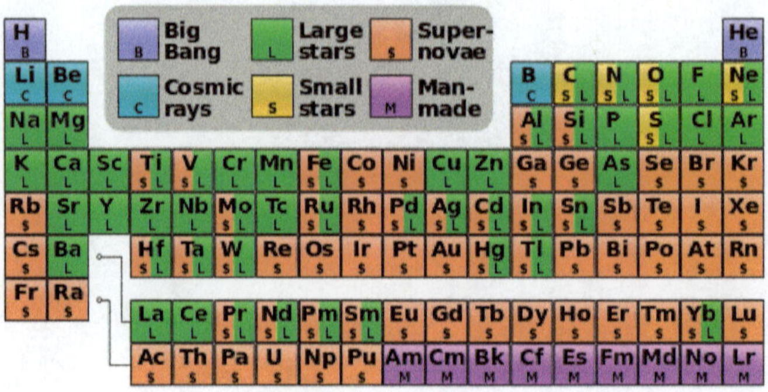

Periodic table showing the cosmogenic origin of each element in the Big Bang, or in large or small stars. Small stars can produce certain elements up to sulfur, by the alpha process. Supernovae are needed to produce "heavy" elements (those beyond iron and nickel) rapidly by neutron buildup, in the r-process. Certain large stars slowly produce other elements heavier than iron, in the s-process; these may then be blown into space in the off-gassing of planetary nebulae

Human technology has produced various additional elements beyond these first 94, with those through atomic number 118 now known.

Abundance

The following graph (note log scale) shows the abundance of elements in our solar system. The table shows the twelve most common elements in our galaxy (estimated spectroscopically), as measured in parts per million, by mass. Nearby galaxies that have evolved along similar lines have a corresponding enrichment of elements heavier than hydrogen and helium. The more distant galaxies are being viewed as they appeared in the past, so their abundances of elements appear closer to the primordial mixture. As physical laws and processes appear common throughout the visible universe, however, scientist expect that these galaxies evolved elements in similar abundance.

The abundance of elements in the Solar System is in keeping with their origin from nucleosynthesis in the Big Bang and a number of progenitor supernova stars. Very abundant hydrogen and helium are products of the Big Bang, but the next three elements are rare since they had little time to form in the Big Bang and are not made in stars (they are, however, produced in small quantities by the breakup of heavier elements in interstellar dust, as a result of impact by cosmic rays). Beginning with carbon, elements are produced in stars by buildup from alpha particles (helium nuclei), resulting in an alternatingly larger abundance of elements with even atomic numbers (these are also more stable). In general, such elements up to iron are made in large stars in the process of becoming supernovas. Iron-56 is particularly common, since it is the most stable element that can easily be made from alpha particles (being a product of decay of radioactive nickel-56, ultimately made from 14 helium nuclei). Elements heavier than iron are made in energy-absorbing processes in large stars, and their abundance in the universe (and on Earth) generally decreases with their atomic number.

The abundance of the chemical elements on Earth varies from air to crust to ocean, and in various types of life. The abundance of elements in Earth's crust differs from that in the Solar system (as seen in the Sun and heavy planets like Jupiter) mainly in selective loss of the very lightest elements (hydrogen and helium) and also volatile neon, carbon (as hydrocarbons), nitrogen and sulfur, as a result of solar heating in the early formation of the solar system. Oxygen, the most abundant Earth element by mass, is retained on Earth by combination with silicon. Aluminum at 8% by mass is more common in the Earth's crust than in the universe and solar system, but the composition of the far more bulky mantle, which has magnesium and iron in place of aluminum (which occurs there only at 2% of mass) more closely mirrors the elemental composition of the solar system, save for the noted loss of volatile elements to space, and loss of iron which has migrated to the Earth's core.

The composition of the human body, by contrast, more closely follows the composition of seawater—save that the human body has additional stores of carbon and nitrogen necessary to form the proteins and nucleic acids, together with phosphorus in the nucleic acids and energy transfer molecule adenosine triphosphate (ATP) that occurs in the cells of all living organisms. Certain kinds of organisms require particular additional elements, for example the magnesium in chlorophyll in green plants, the calcium in mollusc shells, or the iron in the hemoglobin in vertebrate animals' red blood cells.

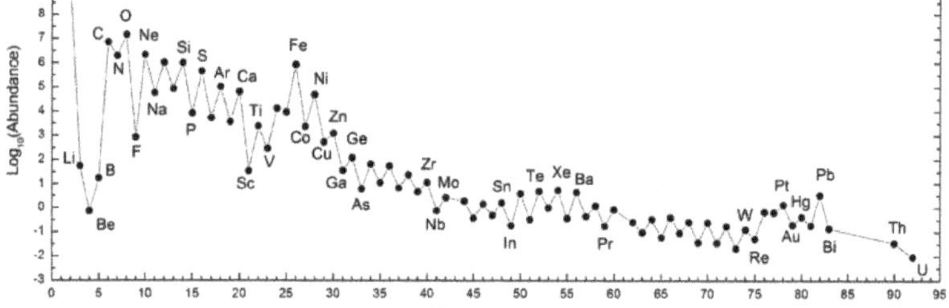

Abundances of the chemical elements in the Solar system. Hydrogen and helium are most common, from the Big Bang. The next three elements (Li, Be, B) are rare because they are poorly synthesized in the Big Bang and also in stars. The two general trends in the remaining stellar-produced elements are: (1) an alternation of abundance in elements as they have even or odd atomic numbers (the Oddo-Harkins rule), and (2) a general decrease in abundance as elements become heavier. Iron is especially common because it represents the minimum energy nuclide that can be made by fusion of helium in supernovae.

Elements in our galaxy	Parts per million by mass
Hydrogen	739,000
Helium	240,000
Oxygen	10,400
Carbon	4,600
Neon	1,340
Iron	1,090
Nitrogen	960
Silicon	650
Magnesium	580
Sulfur	440
Potassium	210
Nickel	100

Nutritional elements in the periodic table

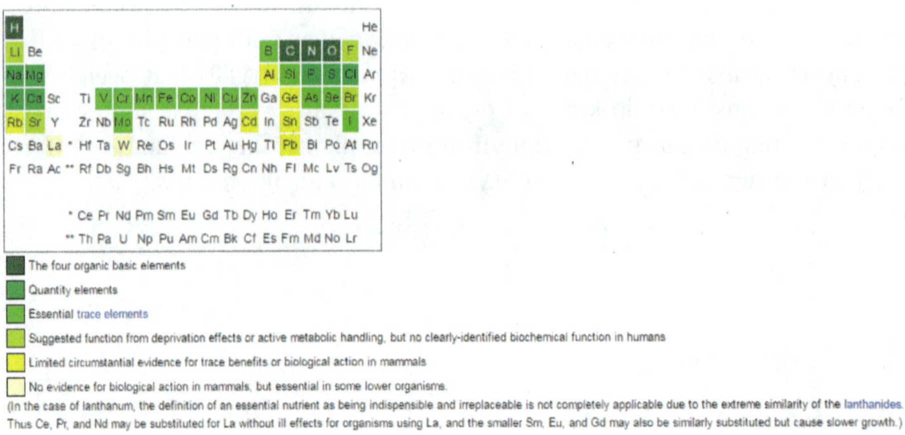

Evolving Definitions

The concept of an "element" as an undivisible substance has developed through three major historical phases: Classical definitions (such as those of the ancient Greeks), chemical definitions, and atomic definitions.

ОПЫТЪ СИСТЕМЫ ЭЛЕМЕНТОВЪ,

ОСНОВАННОЙ НА ИХЪ АТОМНОМЪ ВѢСѢ И ХИМИЧЕСКОМЪ СХОДСТВѢ.

			Ti=50	Zr=90	?=180.
			V=51	Nb=94	Ta=182.
			Cr=52	Mo=96	W=186.
			Mn=55	Rh=104,4	Pt=197,1.
			Fe=56	Ru=104,4	Ir=198.
			Ni=Co=59	Pd=106,6	Os=199.
H=1			Cu=63,4	Ag=108	Hg=200.
	Be=9,4	Mg=24	Zn=65,2	Cd=112	
	B=11	Al=27,3	?=68	Ur=116	Au=197?
	C=12	Si=28	?=70	Sn=118	
	N=14	P=31	As=75	Sb=122	Bi=210?
	O=16	S=32	Se=79,4	Te=128?	
	F=19	Cl=35,5	Br=80	I=127	
Li=7	Na=23	K=39	Rb=85,4	Cs=133	Tl=204.
		Ca=40	Sr=87,6	Ba=137	Pb=207.
		?=45	Ce=92		
		?Er=56	La=94		
		?Yt=60	Di=95		
		?In=75,6	Th=118?		

Mendeleev's 1869 periodic table: An experiment on a system of elements. Based on their atomic weights and chemical similarities.

Classical Definitions

Ancient philosophy posited a set of classical elements to explain observed patterns in nature. These elements originally referred to earth, water, air and fire rather than the chemical elements of modern science.

The term 'elements' (stoicheia) was first used by the Greek philosopher Plato in about 360 BCE in his dialogue Timaeus, which includes a discussion of the composition of inorganic and organic bodies and is a speculative treatise on chemistry. Plato believed the elements introduced a century earlier by Empedocles were composed of small polyhedral forms: tetrahedron (fire), octahedron (air), icosahedron (water), and cube (earth).

Aristotle, c. 350 BCE, also used the term stoicheia and added a fifth element called aether, which formed the heavens. Aristotle defined an element as:

Element – one of those bodies into which other bodies can decompose, and that itself is not capable of being divided into other.

Chemical Definitions

In 1661, Robert Boyle proposed his theory of corpuscularism which favoured the analysis of matter as constituted by irreducible units of matter (atoms) and, choosing to side with neither Aristotle's view of the four elements nor Paracelsus' view of three fundamental elements, left open the question of the number of elements. The first modern list of chemical elements was given in Antoine Lavoisier's 1789 Elements of Chemistry, which contained thirty-three elements, including light and caloric. By 1818, Jöns Jakob Berzelius had determined atomic weights for forty-five of the forty-nine then-accepted elements. Dmitri Mendeleev had sixty-six elements in his periodic table of 1869.

Dmitri Mendeleev

From Boyle until the early 20th century, an element was defined as a pure substance that could not be decomposed into any simpler substance. Put another way, a chemical element cannot be transformed into other chemical elements by chemical processes. Elements during this time were generally distinguished by their atomic weights, a property measurable with fair accuracy by available analytical techniques.

Atomic Definitions

The 1913 discovery by English physicist Henry Moseley that the nuclear charge is the physical basis for an atom's atomic number, further refined when the nature of protons and neutrons became appreciated, eventually led to the current definition of an element based on atomic number (number of protons per atomic nucleus). The use of atomic numbers, rather than atomic weights, to distinguish elements has greater predictive value (since these numbers are integers), and also resolves some ambiguities in the chemistry-based view due to varying properties of isotopes and allotropes within the same element. Currently, IUPAC defines an element to exist if it has isotopes with a lifetime longer than the 10^{-14} seconds it takes the nucleus to form an electronic cloud.

Henry Moseley

By 1914, seventy-two elements were known, all naturally occurring. The remaining naturally occurring elements were discovered or isolated in subsequent decades, and various additional elements have also been produced synthetically, with much of that work pioneered by Glenn T. Seaborg. In 1955, element 101 was discovered and named mendelevium in honor of D.I. Mendeleev, the first to arrange the elements in a periodic manner. Most recently, the synthesis of element 118 was reported in October 2006, and the synthesis of element 117 was reported in April 2010.

Discovery and Recognition of Various Elements

Ten materials familiar to various prehistoric cultures are now known to be chemical elements: Carbon, copper, gold, iron, lead, mercury, silver, sulfur, tin, and zinc. Three additional materials now accepted as elements, arsenic, antimony, and bismuth, were recognized as distinct substances prior to 1500 AD. Phosphorus, cobalt, and platinum were isolated before 1750.

Most of the remaining naturally occurring chemical elements were identified and characterized by 1900, including:

- Such now-familiar industrial materials as aluminium, silicon, nickel, chromium, magnesium, and tungsten
- Reactive metals such as lithium, sodium, potassium, and calcium
- The halogens fluorine, chlorine, bromine, and iodine
- Gases such as hydrogen, oxygen, nitrogen, helium, argon, and neon
- Most of the rare-earth elements, including cerium, lanthanum, gadolinium, and neodymium.
- The more common radioactive elements, including uranium, thorium, radium, and radon

Elements isolated or produced since 1900 include:

- The three remaining undiscovered regularly occurring stable natural elements: hafnium, lutetium, and rhenium
- Plutonium, which was first produced synthetically in 1940 by Glenn T. Seaborg, but is now also known from a few long-persisting natural occurrences
- The three incidentally occurring natural elements (neptunium, promethium, and technetium), which were all first produced synthetically but later discovered in trace amounts in certain geological samples
- Three scarce decay products of uranium or thorium, (astatine, francium, and protactinium), and
- Various synthetic transuranic elements, beginning with americium and curium

Recently Discovered Elements

The first transuranium element (element with atomic number greater than 92) discovered was neptunium in 1940. Since 1999 claims for the discovery of new elements have been considered by the IUPAC/IUPAP Joint Working Party. As of January 2016, all 118 elements have been confirmed as discovered by IUPAC. The discovery of element 112 was acknowledged in 2009, and the name copernicium and the atomic symbol Cn were suggested for it. The name and symbol were officially endorsed by IUPAC on 19 February 2010. The heaviest element that is believed to have been synthesized to date is element 118, oganesson, on 9 October 2006, by the Flerov Laboratory of Nuclear Reactions in Dubna, Russia. Tennessine, element 117 was the latest element claimed to be discovered, in 2009. On 28 November 2016, scientists at the International Union of Pure and Applied Chemistry officially recognized the names for four of the newest chemical elements: Nihonium, Nh, 113; Moscovium, Mc, 115; Tennessine, Ts, 117 and Oganesson, Og, 118.

CHEMICAL REVOLUTION

The chemical revolution, also called the first chemical revolution, was the early modern reformulation of chemistry that culminated in the law of conservation of mass and the oxygen theory of combustion. During the 19th and 20th century, this transformation was credited to the work of the French chemist Antoine Lavoisier (the "father of modern chemistry"). However, recent work on the history of early modern chemistry considers the chemical revolution to consist of gradual changes in chemical theory and practice that emerged over a period of two centuries. The so-called scientific revolution took place during the sixteenth and seventeenth centuries whereas the chemical revolution took place during the seventeenth and eighteenth centuries.

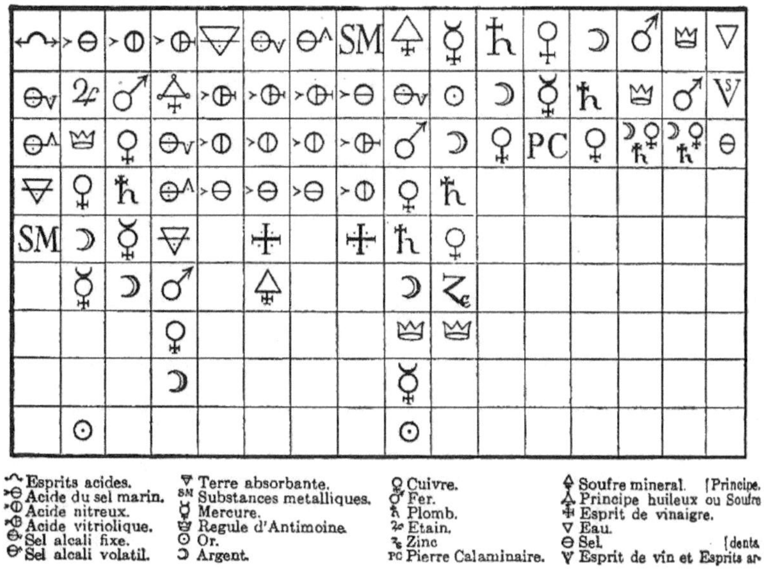

Geoffroy's 1718 Affinity Table: at the head of each column is a chemical species with which all the species below can combine. Some historians have defined this table as being the start of the chemical revolution.

Primary Factors

Several factors led to the first chemical revolution. First, there were the forms of gravimetric analysis that emerged from alchemy and new kinds of instruments that were developed in medical and industrial contexts. In these settings, chemists increasingly challenged hypotheses that had already been presented by the ancient Greeks. For example, chemists began to assert that all structures were composed of more than the four elements of the Greeks or the eight elements of the medieval alchemists. The Irish alchemist, Robert Boyle, laid the foundations for the Chemical Revolution, with his mechanical corpuscular philosophy, which in turn relied heavily on the alchemical corpuscular theory and experimental method dating back to pseudo-Geber.

Other factors included new experimental techniques and the discovery of 'fixed air' (carbon dioxide) by Joseph Black in the middle of the 18th century. This discovery was particularly important because it empirically proved that 'air' did not consist of only one substance and because it established 'gas' as an important experimental substance. Nearer the end of the 18th century, the experiments by Henry Cavendish and Joseph Priestley further proved that air is not an element and is instead composed of several different gases. Lavoisier also translated the names of chemical substance into a new nomenclatural language more appealing to scientists of the nineteenth century. Such changes took place in an atmosphere in which the industrial revolution increased public interest in learning and practicing chemistry. When describing the task of reinventing chemical nomenclature, Lavoisier attempted to harness the new centrality of chemistry by making the rather hyperbolic claim that:

"We must clean house thoroughly, for they have made use of an enigmatical language peculiar to themselves, which in general presents one meaning for the adepts and another meaning for the vulgar, and at the same time contains nothing that is rationally intelligible either for the one or for the other.

Precision Instruments

Much of the reasoning behind Antoine Lavoisier being named the "father of modern chemistry" and the start of the chemical revolution lay in his ability to mathematize the field, pushing chemistry to use the experimental methods utilized in other "more exact sciences." Lavoisier changed the field of chemistry by keeping meticulous balance sheets in his research, attempting to show that through the transformation of chemical species the total amount of substance was conserved. Lavoisier used instrumentation for thermometric and barometric measurements in his experiments, and collaborated with Pierre Simon de Laplace in the invention of the calorimeter, an instrument for measuring heat changes in a reaction. In attempting to dismantle phlogiston theory and implement his own theory of combustion, Lavoisier utilized multiple apparatuses. These included a red-hot iron gun barrel which was designed to have water run through it and decomposed, and an alteration of the apparatus which implemented a pneumatic trough at one end, a thermometer, and a barometer. The precision of his measurements were a requirement in convincing opposition of his theories about water as a compound, with instrumentation designed by himself implemented in his research.

Despite having precise measurements for his work, Lavoisier faced a large amount of opposition in his research. Proponents of phlogiston theory, such as Keir and Priestley, claimed that demonstration of facts was only applicable for raw phenomena, and that interpretation of these facts did not imply accuracy in theories. They stated that Lavoisier was attempting to impose order on observed phenomena, whereas a secondary source of validity would be required to give definitive proof of the composition of water and non-existence of phlogiston.

Antoine Lavoisier

The latter stages of the revolution was fuelled by the 1789 publication of Lavoisier's Traité Élémentaire de Chimie (Elements of Chemistry). Beginning with this publication and others to follow, Lavoisier synthesised the work of others and coined the term "oxygen". Antoine Lavoisier represented the chemical revolution not only in his publications, but also in the way he practiced chemistry. Lavoisier›s work was characterized by his systematic determination of weights and his strong emphasis on precision and accuracy. The law of the conservation of mass, that the weight of matter would be conserved through any reaction, was discovered by Antoine Lavoisier in the late 18th century and represented his careful conduction of research.

Lavoisier also contributed to chemistry a method of understanding combustion and respiration and proof of the composition of water by decomposition into its constituent parts. He explained the theory of combustion, and challenged the phlogiston theory with his views on caloric. The Traité incorporates notions of a "new chemistry" and describes the experiments and reasoning that led to his conclusions. Like Newton›s Principia, which was the high point of the Scientific Revolution, Lavoisier›s Traité can be seen as the culmination of the Chemical Revolution.

Lavoisier›s work was not immediately accepted and it took several decades for it gain momentum. This transition was aided by the work of Jöns Jakob Berzelius, who came up with a simplified shorthand to describe chemical compounds based on John Dalton›s theory of atomic weights. Many people credit Lavoisier and his overthrow of phlogiston theory as the traditional chemical revolution, with Lavoisier marking the beginning of the revolution and John Dalton marking its culmination.

CHEMICAL STRUCTURE

A chemical structure determination includes a chemist's specifying the molecular geometry and, when feasible and necessary, the electronic structure of the target molecule or other solid. Molecular geometry refers to the spatial arrangement of atoms in a molecule and the chemical bonds that hold the atoms together, and can be represented using structural formulae and by molecular models; complete electronic structure descriptions include specifying the occupation of a molecule's molecular orbitals. Structure determination can be applied to a range of targets from very simple molecules (e.g., diatomic oxygen or nitrogen), to very complex ones (e.g., such as of protein or DNA).

Theories of chemical structure were first developed by August Kekule, Archibald Scott Couper, and Aleksandr Butlerov, among others, from about 1858. These theories were first to state that chemical compounds are not a random cluster of atoms and functional groups, but rather had a definite order defined by the valency of the atoms composing the molecule, giving the molecules a three dimensional structure that could be determined or solved.

In determining structures of chemical compounds, one generally aims to obtain, minimally, the pattern and multiplicity of bonding between all atoms in the molecule; when possible, one seeks the three dimensional spatial coordinates of the atoms in the molecule (or other solid). The methods by which one can elucidate the structure of a molecule include spectroscopies such as nuclear magnetic resonance (proton and carbon-13 NMR), various methods of mass spectrometry (to give overall molecular mass, as well as fragment masses), and x-ray crystallography when applicable. The last technique can produce three-dimensional models at atomic-scale resolution, as long as crystals are available. When a molecule has an unpaired electron spin in a functional group of its structure, ENDOR and electron-spin resonance spectroscopes may also be performed. Techniques such as absorption spectroscopy and the vibrational spectroscopies, infrared and Raman, provide, respectively, important supporting information about the numbers and adjacencies of multiple bonds, and about the types of functional groups (whose internal bonding gives vibrational signatures); further inferential studies that give insight into the contributing electronic structure of molecules include cyclic voltammetry and X-ray photoelectron spectroscopy. These latter techniques become all the more important when the molecules contain metal atoms, and when the crystals required by crystallography or the specific atom types that are required by NMR are unavailable to exploit in the structure determination. Finally, more specialized methods such as electron microscopy are also applicable in some cases.

Chemical Nomenclature

A chemical nomenclature is a set of rules to generate systematic names for chemical compounds. The nomenclature used most frequently worldwide is the one created and developed by the International Union of Pure and Applied Chemistry (IUPAC).

The IUPAC's rules for naming organic and inorganic compounds are contained in two publications, known as the Blue Book and the Red Book, respectively. A third publication, known as the Green Book, describes the recommendations for the use of symbols for physical quantities (in association with the IUPAP), while a fourth, the Gold Book, contains the definitions of a large number of technical terms used in chemistry. Similar compendia exist for biochemistry (the White Book, in association with the IUBMB), analytical chemistry (the Orange Book), macromolecular chemistry (the Purple Book) and clinical chemistry (the Silver Book). These "color books" are supplemented by shorter recommendations for specific circumstances that are published periodically in the journal Pure and Applied Chemistry.

Aims of Chemical Nomenclature

The primary function of chemical nomenclature is to ensure that a spoken or written chemical name leaves no ambiguity concerning which chemical compound the name refers to: each chemical name should refer to a single substance. A less important aim is to ensure that each substance has a single name, although a limited number of alternative names is acceptable in some cases.

Preferably, the name also conveys some information about the structure or chemistry of a compound. CAS numbers form an extreme example of names that do not perform this function: each CAS number refers to a single compound but none contain information about the structure.

The form of nomenclature used depends on the audience to which it is addressed. As such, no single correct form exists, but rather there are different forms that are more or less appropriate in different circumstances.

A common name will often suffice to identify a chemical compound in a particular set of circumstances. To be more generally applicable, the name should indicate at least the chemical formula. To be more specific still, the three-dimensional arrangement of the atoms may need to be specified.

In a few specific circumstances (such as the construction of large indices), it becomes necessary to ensure that each compound has a unique name: This requires the addition of extra rules to the standard IUPAC system (the CAS system is the most commonly used in this context), at the expense of having names that are longer and less familiar to most readers. Another system gaining popularity is the International Chemical Identifier (InChI) – which reflects a substance's structure and composition, making it more general than a CAS number.

The IUPAC system is often criticized for the above failures when they become relevant (for example, in differing reactivity of sulfur allotropes, which IUPAC does not distinguish). While IUPAC has a human-readable advantage over CAS numbering, it would be difficult to claim that the IUPAC names for some larger, relevant molecules (such as rapamycin) are human-readable, and so most researchers simply use the informal names.

Differing aims of Chemical Nomenclature and Lexicography

It's generally understood that the aims of lexicography versus chemical nomenclature vary and are to an extent at odds. Dictionaries of words, whether in traditional print or on the web, collect and report the meanings of words as their uses appear and change over time. For web dictionaries with limited or no formal editorial process, definitions—in this case, definitions of chemical names and terms—can change rapidly without concern for the formal or historical meanings. Chemical nomenclature on the other hand (with IUPAC nomenclature as the best example) is necessarily more restrictive: It aims to standardize communication and practice so that, when a chemical term is used it has a fixed meaning relating to chemical structure, thereby giving insights into chemical properties and derived molecular functions. These differing aims can have profound effects on valid understanding in chemistry, especially with regard to chemical classes that have achieved mass attention. Examples of the impact of these can be seen in considering the examples of:

- resveratrol, a single compound clearly defined by this common name, but that can be confused, popularly, with its cis-isomer,
- omega-3 fatty acids, a reasonably well-defined chemical structure class that is nevertheless broad as a result of its formal definition, and
- polyphenols, a fairly broad structural class with a formal definition, but where mistranslations and general misuse of the term relative to the formal definition has led to serious usage errors, and so ambiguity in the relationship between structure and activity (SAR).

The rapid pace at which meanings can change on the web, in particular for chemical compounds with perceived health benefits, rightly or wrongly ascribed, complicates the matter of maintaining a sound nomenclature (and so access to SAR understanding). A further discussion with specific examples appears in the article on polyphenols, where differing definitions are in use, and there are various, further web definitions and common uses of the word at odds with any accepted chemical nomenclature connecting polyphenol structure and bioactivity).

Types of Nomenclature

Organic Chemistry

- Substitutive name
- Functional class name, also known as a radicofunctional name

- Conjunctive name
- Additive name
- Subtractive name
- Multiplicative name
- Fusion name
- Hantzsch–Widman name
- Replacement name

IUPAC NOMENCLATURE OF INORGANIC CHEMISTRY

In chemical nomenclature, the IUPAC nomenclature of inorganic chemistry is a systematic method of naming inorganic chemical compounds, as recommended by the International Union of Pure and Applied Chemistry (IUPAC). It is published in Nomenclature of Inorganic Chemistry (which is informally called the Red Book). Ideally, every inorganic compound should have a name from which an unambiguous formula can be determined. There is also an IUPAC nomenclature of organic chemistry.

System

The names "caffeine" and "3,7-dihydro-1,3,7-trimethyl-1H-purine-2,6-dione" both signify the same chemical. The systematic name encodes the structure and composition of the caffeine molecule in some detail, and provides an unambiguous reference to this compound, whereas the name "caffeine" just names it. These advantages make the systematic name far superior to the common name when absolute clarity and precision are required. However, for the sake of brevity, even professional chemists will use the non-systematic name almost all of the time, because caffeine is a well-known common chemical with a unique structure. Similarly, H2O is most often simply called water in English, though other chemical names do exist.

- Single atom anions are named with an -ide suffix: for example, H− is hydride.
- Compounds with a positive ion (cation): The name of the compound is simply the cation's name (usually the same as the element's), followed by the anion. For example, NaCl is sodium chloride, and CaF_2 is calcium fluoride.
- Cations which have taken on more than one positive charge are labeled with Roman numerals in parentheses. For example, Cu^+ is copper(I), Cu^{2+} is copper(II). An older, deprecated notation is to append -ous or -ic to the root of the Latin name to name ions with a lesser or greater charge. Under this naming convention, Cu^+ is cuprous and Cu^{2+} is cupric. For naming metal complexes see the page on complex (chemistry).
- Oxyanions (polyatomic anions containing oxygen) are named with -ite or -ate, for a lesser or greater quantity of oxygen, respectively. For example, NO_2^- is nitrite, while NO_3^- is nitrate. If four oxyanions are possible, the prefixes *hypo-* and *per-* are used: hypochlorite is ClO^-, perchlorate is ClO_4^-.
- The prefix bi- is a deprecated way of indicating the presence of a single

hydrogen ion, as in “sodium bicarbonate” ($NaHCO_3$). The modern method specifically names the hydrogen atom. Thus, $NaHCO_3$ would be pronounced sodium hydrogen carbonate.

Positively charged ions are called cations and negatively charged ions are called anions. The cation is always named first. Ions can be metals or polyatomic ions. Therefore the name of the metal or positive polyatomic ion is followed by the name of the non-metal or negative polyatomic ion. The positive ion retains its element name whereas for a single non-metal anion the ending is changed to -ide.

Example: sodium chloride, potassium oxide, or calcium carbonate.

When the metal has more than one possible ionic charge or oxidation number the name becomes ambiguous. In these cases the oxidation number (the same as the charge) of the metal ion is represented by a Roman numeral in parentheses immediately following the metal ion name. For example in uranium(VI) fluoride the oxidation number of uranium is 6. Another example is the iron oxides. F_eO is iron(II) oxide and Fe_2O_3 is iron(III) oxide.

An older system used prefixes and suffixes to indicate the oxidation number, according to the following scheme:

Oxidation state	Cations and acids	Anions
Lowest	*hypo- -ous*	*hypo- -ite*
	-ous	*-ite*
	-ic	*-ate*
	per- -ic	*per- -ate*
Highest	*hyper- -ic*	*hyper- -ate*

Thus the four oxyacids of chlorine are called hypochlorous acid (HOCl), chlorous acid (HOClO), chloric acid ($HOClO_2$) and perchloric acid ($HOClO_3$), and their respective conjugate bases are the hypochlorite, chlorite, chlorate and perchlorate ions. This system has partially fallen out of use, but survives in the common names of many chemical compounds: the modern literature contains few references to “ferric chloride” (instead calling it “iron(III) chloride”), but names like “potassium permanganate” (instead of “potassium manganate(VII)”) and “sulfuric acid” abound.

REFERENCES

1. Whitten, Kenneth W.; Davis, Raymond E.; Peck, M. Larry (2000), General Chemistry (6th ed.), Fort Worth, TX: Saunders College Publishing/Harcourt College Publishers, ISBN 978-0-03-072373-5
2. Brown, Theodore L.; LeMay, H. Eugene; Bursten, Bruce E.; Murphy, Catherine J.; Woodward, Patrick (2009), Chemistry: The Central Science, AP Edition (11th ed.), Upper Saddle River, NJ: Pearson/Prentice Hall, pp. 5–6, ISBN 0-13-236489-1
3. Hill, John W.; Petrucci, Ralph H.; McCreary, Terry W.; Perry, Scott S. (2005), General Chemistry (4th ed.), Upper Saddle River, NJ: Pearson/Prentice Hall, p. 6, ISBN 978-0-13-140283-6

4. Wilbraham, Antony; Matta, Michael; Staley, Dennis; Waterman, Edward (2002), Chemistry (1st ed.), Upper Saddle River, NJ: Pearson/Prentice Hall, p. 36, ISBN 0-13-251210-6

5. Manchester, F. D.; San-Martin, A.; Pitre, J. M. (1994). "The H-Pd (hydrogen-palladium) System". Journal of Phase Equilibria. 15: 62. doi:10.1007/BF02667685. Phase diagram for Palladium-Hydrogen System

6. Atkins, Peter; Jones, Loretta (2004). Chemical Principles: The Quest for Insight.

7. IUPAC (ed.). "chemical element". International Union of Pure and Applied Chemistry. doi:10.1351/goldbook.C01022.

8. Los Alamos National Laboratory (2011). "Periodic Table of Elements: Oxygen". Los Alamos, New Mexico: Los Alamos National Security, LLC. Retrieved 7 May 2011.

9. Oerter, Robert (2006). The Theory of Almost Everything: The Standard Model, the Unsung Triumph of Modern Physics. Penguin. p. 223. ISBN 978-0-452-28786-0.

10. E. M. Burbidge; G. R. Burbidge; W. A. Fowler; F. Hoyle (1957). "Synthesis of the Elements in Stars". Reviews of Modern Physics. 29 (4): 547–650. Bibcode:1957RvMP...29..547B. doi:10.1103/RevModPhys.29.547.

11. See the timeline on p.10 in Oganessian, Yu. Ts.; Utyonkov, V.; Lobanov, Yu.; Abdullin, F.; Polyakov, A.; Sagaidak, R.; Shirokovsky, I.; Tsyganov, Yu.; et al. (2006). "Evidence for Dark Matter" (PDF). Physical Review C. 74 (4): 044602. Bibcode:2006PhRvC..74d4602O. doi:10.1103/PhysRevC.74.044602.

12. lbl.gov (2005). "The Universe Adventure Hydrogen and Helium". Lawrence Berkeley National Laboratory U.S. Department of Energy.

13. astro.soton.ac.uk (January 3, 2001). "Formation of the light elements". University of Southampton.

14. foothill.edu (October 18, 2006). "How Stars Make Energy and New Elements"(PDF). Foothill College.

15. Dumé, B (23 April 2003). "Bismuth breaks half-life record for alpha decay". Physicsworld.com. Bristol, England: Institute of Physics. Retrieved 14 July 2015.

16. de Marcillac, P; Coron, N; Dambier, G; Leblanc, J; Moalic, J-P (2003). "Experimental detection of alpha-particles from the radioactive decay of natural bismuth". Nature. 422 (6934): 876–8. Bibcode:2003Natur.422..876D. doi:10.1038/nature01541. PMID 12712201.

17. Sanderson, K (17 October 2006). "Heaviest element made – again". Nature News. doi:10.1038/news061016-4.

18. Schewe, P; Stein, B (17 October 2000). "Elements 116 and 118 Are Discovered". Physics News Update. American Institute of Physics. Retrieved 19 October 2006.

19. United States Environmental Protection Agency. "Technetium-99". epa.gov. Retrieved 26 February 2013.

20. Harvard–Smithsonian Center for Astrophysics. "ORIGIN OF HEAVY ELEMENTS". cfa.harvard.edu. Retrieved 26 February 2013.

21. "ATOMIC NUMBER AND MASS NUMBERS". ndt-ed.org. Retrieved 17 February 2013.
22. periodic.lanl.gov. "PERIODIC TABLE OF ELEMENTS: LANL Carbon". Los Alamos National Laboratory.
23. Katsuya Yamada. "Atomic mass, isotopes, and mass number." (PDF). Los Angeles Pierce College.
24. "Pure element". European Nuclear Society.
25. Wilford, JN (14 January 1992). "Hubble Observations Bring Some Surprises". New York Times.
26. Wright, EL (12 September 2004). "Big Bang Nucleosynthesis". UCLA, Division of Astronomy. Retrieved 22 February 2007.
27. Wallerstein, George; Iben, Icko; Parker, Peter; Boesgaard, Ann; Hale, Gerald; Champagne, Arthur; Barnes, Charles; Käppeler, Franz; et al. (1999). "Synthesis of the elements in stars: forty years of progress" (PDF). Reviews of Modern Physics. 69 (4): 995–1084. Bibcode:1997RvMP...69..995W. doi:10.1103/RevModPhys.69.995.
28. Earnshaw, A; Greenwood, N (1997). Chemistry of the Elements (2nd ed.). Butterworth-Heinemann.
29. Croswell, K (1996). Alchemy of the Heavens. Anchor. ISBN 0-385-47214-5.
30. Plato (2008). Timaeus. Forgotten Books. p. 45. ISBN 978-1-60620-018-6.
31. Hillar, M (2004). "The Problem of the Soul in Aristotle's De anima". NASA/WMAP. Retrieved 10 August 2006.
32. Partington, JR (1937). A Short History of Chemistry. New York: Dover Publications. ISBN 0-486-65977-1.
33. Boyle, R (1661). The Sceptical Chymist. London. ISBN 0-922802-90-4.
34. Lavoisier, AL (1790). Elements of chemistry translated by Robert Kerr. Edinburgh. pp. 175–6. ISBN 978-0-415-17914-0.
35. Transactinide-2. www.kernchemie.de
36. Carey, GW (1914). The Chemistry of Human Life. Los Angeles. ISBN 0-7661-2840-7.
37. Glanz, J (6 April 2010). "Scientists Discover Heavy New Element". New York Times.
38. "IUPAC Announces Start of the Name Approval Process for the Element of Atomic Number 112" (PDF). IUPAC. 20 July 2009. Retrieved 27 August 2009.
39. "IUPAC (International Union of Pure and Applied Chemistry): Element 112 is Named Copernicium". IUPAC. 20 February 2010. Archived from the original on 24 February 2010.
40. Oganessian, Yu. Ts.; Utyonkov, V.; Lobanov, Yu.; Abdullin, F.; Polyakov, A.; Sagaidak, R.; Shirokovsky, I.; Tsyganov, Yu.; et al. (2006). "Evidence for Dark Matter" (PDF). Physical Review C. 74 (4): 044602. Bibcode:2006PhRvC..74d4602O. doi:10.1103/PhysRevC.74.044602.

41. Greiner, W. "Recommendations" (PDF). 31st meeting, PAC for Nuclear Physics. Joint Institute for Nuclear Research.
42. Staff (30 November 2016). "IUPAC Announces the Names of the Elements 113, 115, 117, and 118". IUPAC. Retrieved 1 December 2016.
43. St. Fleur, Nicholas (1 December 2016). "Four New Names Officially Added to the Periodic Table of Elements". New York Times. Retrieved 1 December 2016.
44. Kim, Mi Gyung (2003). Affinity, That Elusive Dream: A Genealogy of the Chemical Revolution. MIT Press. ISBN 978-0-262-11273-4.
45. The First Chemical Revolution Archived April 26, 2009, at the Wayback Machine. – the Instrument Project, The College of Wooster
46. Matthew Daniel Eddy, Seymour Mauskopf, and William R. Newman (2014). "An Introduction to Chemical Knowledge in the Early Modern World". Osiris. 29: 1–15. doi:10.1086/678110.
47. Matthew Daniel Eddy, Seymour Mauskopf and William R. Newman (Eds.) (2014). Chemical Knowledge in the Early Modern World. Chicago: University of Chicago Press.
48. Ursula Klein (July 2007). "Styles of Experimentation and Alchemical Matter Theory in the Scientific Revolution". Metascience. Springer. 16 (2): 247–256 . doi:10.1007/s11016-007-9095-8. ISSN 1467-9981.
49. Jaffe, B. (1976). Crucibles: The Story of Chemistry from Alchemy to Nuclear Fission (4th ed.). New York: Dover Publications. ISBN 978-0-486-23342-0.
50. Jan Golinski, "Precision instruments and the demonstrative order of proof in Lavoisier's chemistry," Osiris, 2nd Series (1994): 30-47.
51. Levere, Trevor (2001). Transforming Matter. Baltimore, Maryland: The Johns Hopkins University Press. ISBN 0-8018-6610-3.
52. Eddy, Matthew Daniel (2008). The Language of Mineralogy: John Walker, Chemistry and the Edinburgh Medical School 1750-1800. Ashgate.
53. Antoine-Laurent Lavoisier, Elements of Chemistry, trans. Robert Kerr (Edinburgh, 1790; facs. reprint New York: Dover, 1965), pp. xv-xvi.
54. Dear, Peter (2006). The Intelligibility of Nature. The University of Chicago Press. pp. 74–75.
55. Duveen, Denis; Klickstein, Herbert (Sep 1954). "The Introduction of Lavoisier's Chemical Nomenclature into America". Isis. 45 (3).
56. Nomenclature of Organic Chemistry
57. International Union of Pure and Applied Chemistry (2005). Nomenclature of Inorganic Chemistry (IUPAC Recommendations 2005). Cambridge (UK): RSC–IUPAC. ISBN 0-85404-438-8. Electronic version..
58. International Union of Pure and Applied Chemistry (1993). Quantities, Units and Symbols in Physical Chemistry, 2nd edition, Oxford: Blackwell Science. ISBN 0-632-03583-8. Electronic version..

59. Compendium of Chemical Terminology, IMPACT Recommendations (2nd Ed.), Oxford:Blackwell Scientific Publications. (1997)
60. Biochemical Nomenclature and Related Documents, London:Portland Press, 1992.
61. International Union of Pure and Applied Chemistry (1998). Compendium of Analytical Nomenclature (definitive rules 1997, 3rd. ed.). Oxford: Blackwell Science. ISBN 0-86542-6155. .
62. Compendium of Macromolecular Nomenclature, Oxford:Blackwell Scientific Publications, 1991.
63. Compendium of Terminology and Nomenclature of Properties in Clinical Laboratory Sciences, IMPACT Recommendations 1995, Oxford: Blackwell Science, ISBN 0-86542-612-0.
64. Guyton de Morveau, L. B. (1782), J. Phys., 19: 310 Missing or empty |title=(help).
65. Guyton de Morveau, L. B.; Lavoisier, A. L.; Berthollet, C. L.; Fourcroy, A. F. de (1787), Méthode de Nomenclature Chimique, Paris: Cuchet, archived from the original on 2011-07-21.
66. Lavoisier, A. L. (1801), Traité Élémentaire de Chimie (3e ed.), Paris: Deterville.
67. Berzelius, J. J. (1811), J. Phys., 73: 248 Missing or empty |title= (help).
68. Wisniak, Jaime (2000), "Jöns Jacob Berzelius A Guide to the Perplexed Chemist", Chem. Educator, 5 (6): 343–50, doi:10.1007/s00897000430a.
69. "Congrès de nomenclature chimique, Genève 1892", Bull. Soc. Chim. Paris, Ser. 3, 7: xiii–xxiv, 1892.
70.

CHAPTER 5

Chemical Bond

Chapter Contents

- **Introduction**
- **Overview Of Main Types Of Chemical Bonds**
- **History**
- **Bonds In Chemical Formulas**
- **Strong Chemical Bonds**
- **Intermolecular Bonding**
- **Theories Of Chemical Bonding**

Chemical bonds are atoms that are held together within molecules or crystals. They form as the result of atoms sharing electrons. They also form from the complete transfer of electrons. The three main types of bonds are ionic, covalent, and polar covalent. Hydrogen bonds are often included under the heading of chemical bonds as well.

INTRODUCTION

A chemical bond is a lasting attraction between atoms that enables the formation of chemical compounds. The bond may result from the electrostatic force of attraction between atoms with opposite charges, or through the sharing of electrons as in the covalent bonds. The strength of chemical bonds varies considerably; there are "strong bonds" such as covalent or ionic bonds and "weak bonds" such as Dipole-dipole interaction, the London dispersion force and hydrogen bonding.

Since opposite charges attract via a simple electromagnetic force, the negatively charged electrons that are orbiting the nucleus and the positively charged protons in the nucleus attract each other. An electron positioned between two nuclei will be attracted to both of them, and the nuclei will be attracted toward electrons in this position. This attraction constitutes the chemical bond. Due to the matter wave nature of electrons and their smaller mass, they must occupy a much larger amount of volume compared with the nuclei, and this volume occupied by the electrons keeps the atomic nuclei relatively far apart, as compared with the size of the nuclei themselves. This phenomenon limits the distance between nuclei and atoms in a bond.

In general, strong chemical bonding is associated with the sharing or transfer of electrons between the participating atoms. The atoms in molecules, crystals, metals and diatomic gases—indeed most of the physical environment around us—are held together by chemical bonds, which dictate the structure and the bulk properties of matter.

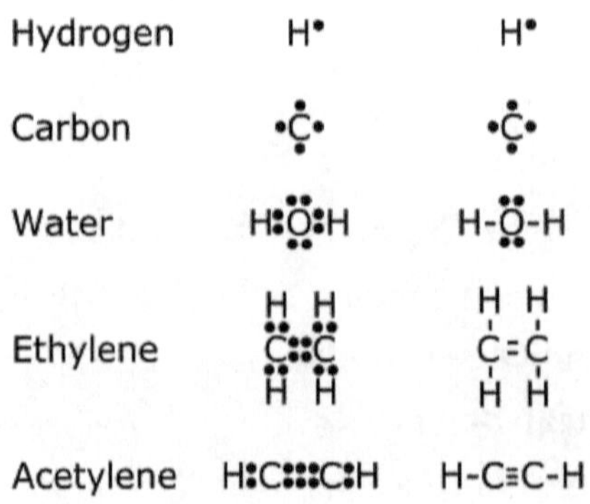

Examples of Lewis dot-style representations of chemical bonds between carbon (C), hydrogen (H), and oxygen (O). Lewis dot diagrams were an early attempt to describe chemical bonding and are still widely used today.

All bonds can be explained by quantum theory, but, in practice, simplification rules allow chemists to predict the strength, directionality, and polarity of bonds. The octet rule and VSEPR theory are two examples. More sophisticated theories are valence bond theory which includes orbital hybridization and resonance, and molecular orbital theory which includes linear combination of atomic orbitals and ligand field theory. Electrostatics are used to describe bond polarities and the effects they have on chemical substances.

OVERVIEW OF MAIN TYPES OF CHEMICAL BONDS

A chemical bond is an attraction between atoms. This attraction may be seen as the result of different behaviors of the outermost or valence electrons of atoms. These behaviors merge into each other seamlessly in various circumstances, so that there is no clear line to be drawn between them. However it remains useful and customary to differentiate between different types of bond, which result in different properties of condensed matter.

In the simplest view of a covalent bond, one or more electrons (often a pair of electrons) are drawn into the space between the two atomic nuclei. Energy is released by bond formation. This is not as a reduction in potential energy, because the attraction of the two electrons to the two protons is offset by the electron-electron and proton-proton repulsions. Instead, the release of energy (and hence stability of the bond) arises from the reduction in kinetic energy due to the electrons being in a more spatially distributed (i.e. longer de Broglie wavelength) orbital compared with each electron being confined closer to its respective nucleus. These bonds exist between two particular identifiable atoms and have a direction in space, allowing them to be shown as single connecting lines between atoms in drawings, or modeled as sticks between spheres in models.

In a polar covalent bond, one or more electrons are unequally shared between two nuclei. Covalent bonds often result in the formation of small collections of better-connected atoms called molecules, which in solids and liquids are bound to other molecules by forces that are often much weaker than the covalent bonds that hold the molecules internally together. Such weak intermolecular bonds give organic molecular substances, such as waxes and oils, their soft bulk character, and their low melting points (in liquids, molecules must cease most structured or oriented contact with each other). When covalent bonds link long chains of atoms in large molecules, however (as in polymers such as nylon), or when covalent bonds extend in networks through solids that are not composed of discrete molecules (such as diamond or quartz or the silicate minerals in many types of rock) then the structures that result may be both strong and tough, at least in the direction oriented correctly with networks of covalent bonds. Also, the melting points of such covalent polymers and networks increase greatly.

In a simplified view of an ionic bond, the bonding electron is not shared at all, but transferred. In this type of bond, the outer atomic orbital of one atom has a vacancy which allows the addition of one or more electrons. These newly added electrons potentially occupy a lower energy-state (effectively closer to more nuclear charge) than they experience in a different atom. Thus, one nucleus offers a more tightly bound position to an electron than does another nucleus, with the result that one atom may transfer an electron to the other. This transfer causes one atom to assume a net positive charge, and the other to assume a net negative charge. The bond then results from electrostatic attraction between atoms and the atoms become positive or negatively charged ions. Ionic bonds may be seen as extreme examples of polarization in covalent bonds. Often, such bonds have no particular orientation in space, since they result from equal electrostatic attraction of each ion to all ions around them. Ionic bonds are strong (and thus ionic substances require high temperatures to melt) but also brittle, since the forces between ions are short-range and do not easily bridge cracks and fractures. This type of bond gives rise to the physical characteristics of crystals of classic mineral salts, such as table salt.

A less often mentioned type of bonding is metallic bonding. In this type of bonding, each atom in a metal donates one or more electrons to a "sea" of electrons that reside between many metal atoms. In this sea, each electron is free (by virtue of its wave nature) to be associated with a great many atoms at once. The bond results because the metal atoms become somewhat positively charged due to loss of their electrons while the electrons remain attracted to many atoms, without being part of any given atom. Metallic bonding may be seen as an extreme example of delocalization of electrons over a large system of covalent bonds, in which every atom participates. This type of bonding is often very strong (resulting in the tensile strength of metals). However, metallic bonding is more collective in nature than other types, and so they allow metal crystals to more easily deform, because they are composed of atoms attracted to each other, but not in any particularly-oriented ways. This results in the malleability of metals. The sea of electrons in metallic bonding causes the characteristically good electrical and thermal conductivity of metals, and also their "shiny" reflection of most frequencies of white light.

HISTORY

Early speculations about the nature of the chemical bond, from as early as the 12th century, supposed that certain types of chemical species were joined by a type of chemical affinity. In 1704, Sir Isaac Newton famously outlined his atomic bonding theory, in "Query 31" of his Opticks, whereby atoms attach to each other by some "force". Specifically, after acknowledging the various popular theories in vogue at the time, of how atoms were reasoned to attach to each other, i.e. "hooked atoms", "glued together by rest", or "stuck together by conspiring motions", Newton states that he would rather infer from their cohesion, that "particles attract one another by some force, which in immediate contact is exceedingly strong, at small distances performs the chemical operations, and reaches not far from the particles with any sensible effect."

In 1819, on the heels of the invention of the voltaic pile, Jöns Jakob Berzelius developed a theory of chemical combination stressing the electronegative and electropositive characters of the combining atoms. By the mid 19th century, Edward Frankland, F.A. Kekulé, A.S. Couper, Alexander Butlerov, and Hermann Kolbe, building on the theory of radicals, developed the theory of valency, originally called "combining power", in which compounds were joined owing to an attraction of positive and negative poles. In 1916, chemist Gilbert N. Lewis developed the concept of the electron-pair bond, in which two atoms may share one to six electrons, thus forming the single electron bond, a single bond, a double bond, or a triple bond; in Lewis's own words, "An electron may form a part of the shell of two different atoms and cannot be said to belong to either one exclusively."

That same year, Walther Kossel put forward a theory similar to Lewis' only his model assumed complete transfers of electrons between atoms, and was thus a model of ionic bonding. Both Lewis and Kossel structured their bonding models on that of Abegg's rule (1904).

In 1927, the first mathematically complete quantum description of a simple chemical bond, i.e. that produced by one electron in the hydrogen molecular ion, H2+, was derived by the Danish physicist Oyvind Burrau. This work showed that the quantum approach to chemical bonds could be fundamentally and quantitatively correct, but the mathematical methods used could not be extended to molecules containing more than one electron. A more practical, albeit less quantitative, approach was put forward in the same year by Walter Heitler and Fritz London. The Heitler-London method forms the basis of what is now called valence bond theory. In 1929, the linear combination of atomic orbitals molecular orbital method (LCAO) approximation was introduced by Sir John Lennard-Jones, who also suggested methods to derive electronic structures of molecules of F2 (fluorine) and O2 (oxygen) molecules, from basic quantum principles. This molecular orbital theory represented a covalent bond as an orbital formed by combining the quantum mechanical Schrödinger atomic orbitals which had been hypothesized for electrons in single atoms. The equations for bonding electrons in multi-electron atoms could not be solved to mathematical perfection (i.e., analytically), but approximations for them still gave many good qualitative predictions and results. Most quantitative calculations in modern quantum chemistry use either valence bond or molecular orbital theory as a starting point, although a third approach, density functional theory, has become increasingly popular in recent years.

In 1933, H. H. James and A. S. Coolidge carried out a calculation on the dihydrogen molecule that, unlike all previous calculation which used functions only of the distance of the electron from the atomic nucleus, used functions which also explicitly added the distance between the two electrons. With up to 13 adjustable parameters they obtained a result very close to the experimental result for the dissociation energy. Later extensions have used up to 54 parameters and gave excellent agreement with experiments. This calculation convinced the scientific community that quantum theory could give agreement with experiment. However this approach has none of the physical pictures of the valence bond and molecular orbital theories and is difficult to extend to larger molecules.

BONDS IN CHEMICAL FORMULAS

Because atoms and molecules are three-dimensional, it is difficult to use a single method to indicate orbitals and bonds. In molecular formulas the chemical bonds (binding orbitals) between atoms are indicated in different ways depending on the type of discussion. Sometimes, some details are neglected. For example, in organic chemistry one is sometimes concerned only with the functional group of the molecule. Thus, the molecular formula of ethanol may be written in conformational form, three-dimensional form, full two-dimensional form (indicating every bond with no three-dimensional directions), compressed two-dimensional form (CH_3–CH_2–OH), by separating the functional group from another part of the molecule (C_2H_5OH), or by its atomic constituents (C_2H_6O), according to what is discussed. Sometimes, even the non-bonding valence shell electrons (with the two-dimensional approximate directions) are marked, e.g. for elemental carbon .'C'. Some chemists may also mark the respective orbitals, e.g. the hypothetical ethene−4 anion (\/C=C/\ −4) indicating the possibility of bond formation.

STRONG CHEMICAL BONDS

Strong chemical bonds are the intramolecular forces which hold atoms together in molecules. A strong chemical bond is formed from the transfer or sharing of electrons between atomic centers and relies on the electrostatic attraction between the protons in nuclei and the electrons in the orbitals.

The types of strong bond differ due to the difference in electronegativity of the constituent elements. A large difference in electronegativity leads to more polar (ionic) character in the bond.

Ionic Bonding

Ionic bonding is a type of electrostatic interaction between atoms which have a large electronegativity difference. There is no precise value that distinguishes ionic from covalent bonding, but a difference of electronegativity of over 1.7 is likely to be ionic, and a difference of less than 1.7 is likely to be covalent. Ionic bonding leads to separate positive and negative ions. Ionic charges are commonly between −3e to +3e. Ionic bonding commonly occurs in metal salts such as sodium chloride (table salt). A typical feature of ionic bonds is that the species form into ionic crystals, in which no ion is specifically paired with any single other ion, in a specific directional bond. Rather, each species of ion is surrounded by ions of the opposite charge, and the spacing between it and each of the oppositely charged ions near it, is the same for all surrounding atoms of the same type. It is thus no longer possible to associate an ion with any specific other single ionized atom near it. This is a situation unlike that in covalent crystals, where covalent bonds between specific atoms are still discernible from the shorter distances between them, as measured via such techniques as X-ray diffraction.

Typical bond lengths in pm and bond energies in kJ/mol. Bond lengths can be converted to Å by division by 100 (1 Å = 100 pm). Data taken from University of Waterloo.

Bond	Length (pm)	Energy (kJ/mol)
H — Hydrogen		
H–H	74	436
H–O	96	366
H–F	92	568
H–Cl	127	432
C — Carbon		
C–H	109	413
C–C	154	348
C–C=	151	
=C–C≡	147	
=C–C=	148	
C=C	134	614
C≡C	120	839
C–N	147	308
C–O	143	360
C–F	134	488
C–Cl	177	330
N — Nitrogen		
N–H	101	391
N–N	145	170
N≡N	110	945
O — Oxygen		
O–O	148	145
O=O	121	498
F, Cl, Br, I — Halogens		
F–F	142	158
Cl–Cl	199	243
Br–H	141	366
Br–Br	228	193
I–H	161	298
I–I	267	151

Ionic crystals may contain a mixture of covalent and ionic species, as for example salts of complex acids, such as sodium cyanide, NaCN. X-ray diffraction shows that in NaCN, for example, the bonds between sodium cations (Na^+) and the cyanide anions (CN^-) are ionic, with no sodium ion associated with any particular cyanide. However, the bonds between C and N atoms in cyanide are of the covalent type, making each of the carbon and nitrogen associated with just one of its opposite type, to which it is physically much closer than it is to other carbons or nitrogens in a sodium cyanide crystal.

When such crystals are melted into liquids, the ionic bonds are broken first because they are non-directional and allow the charged species to move freely. Similarly, when such salts dissolve into water, the ionic bonds are typically broken by the interaction with water, but the covalent bonds continue to hold. For example, in solution, the cyanide ions, still bound together as single CN^- ions, move independently through the solution, as do sodium ions, as Na^+. In water, charged ions move apart because each of them are more strongly attracted to a number of water molecules, than to each other. The attraction between ions and water molecules in such solutions is due to a type of weak dipole-dipole type chemical bond. In melted ionic compounds, the ions continue to be attracted to each other, but not in any ordered or crystalline way.

Covalent Bond

Covalent bonding is a common type of bonding, in which two atoms share two valence electrons, one from each of the atoms. In nonpolar covalent bonds, the electronegativity difference between the bonded atoms is small, typically 0 to 0.3. Bonds within most organic compounds are described as covalent. The figure shows methane (CH_4), in which each hydrogen forms a covalent bond with the carbon. See sigma bonds and pi bonds for LCAO-description of such bonding.

Molecules which are formed primarily from non-polar covalent bonds are often immiscible in water or other polar solvents, but much more soluble in non-polar solvents such as hexane.

A polar covalent bond is a covalent bond with a significant ionic character. This means that the two shared electrons are closer to one of the atoms than the other, creating an imbalance of charge. Such bonds occur between two atoms with moderately different electronegativities and give rise to dipole-dipole interactions. The electronegativity difference between the two atoms in these bonds is 0.3 to 1.7.

Single and Multiple Bonds

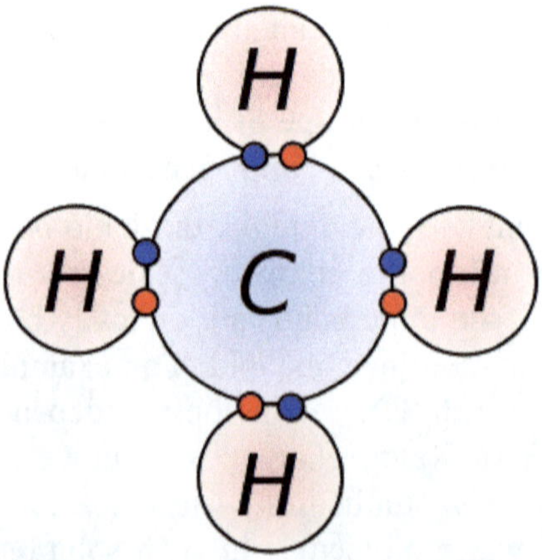

Nonpolar covalent bonds in methane (CH4). The Lewis structure shows electrons shared between C and H atoms.

A single bond between two atoms corresponds to the sharing of one pair of electrons. The electron density of these two bonding electrons is concentrated in the region between the two atoms, which is the defining quality of a sigma bond.

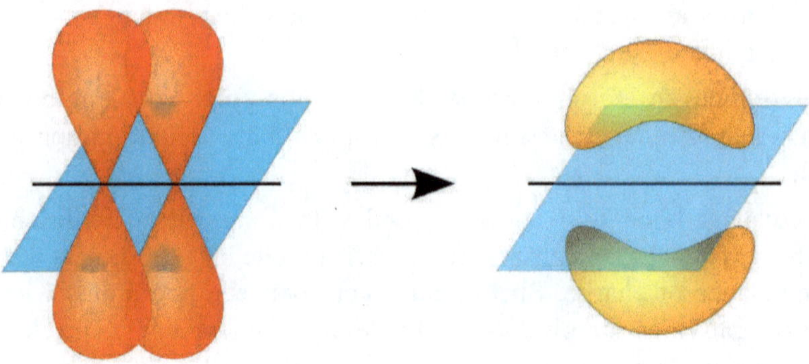

Two p-orbitals forming a pi-bond.

A double bond between two atoms is formed by the sharing of two pairs of electrons, one in a sigma bond and one in a pi bond, with electron density concentrated on two opposite sides of the internuclear axis. A triple bond consists of three shared electron pairs, forming one sigma and two pi bonds.

Quadruple and higher bonds are very rare and occur only between certain transition metal atoms.

Coordinate Covalent Bond (Dipolar Bond)

A coordinate covalent bond is a covalent bond in which the two shared bonding electrons are from the same one of the atoms involved in the bond. For example, boron trifluoride (BF3) and ammonia (NH3) from an adduct or coordination complex F3B←NH3 with a B–N bond in which a lone pair of electrons on N is shared with an empty atomic orbital on B. BF3 with an empty orbital is described as an electron pair acceptor or Lewis acid, while NH3 with a lone pair which can be shared is described as an electron-pair donor or Lewis base. The electrons are shared roughly equally between the atoms in contrast to ionic bonding. Such bonding is shown by an arrow pointing to the Lewis acid.

Transition metal complexes are generally bound by coordinate covalent bonds. For example, the ion Ag^+ reacts as a Lewis acid with two molecules of the Lewis base NH_3 to form the complex ion $Ag(NH_3)_2^+$, which has two Ag←N coordinate covalent bonds.

Metallic Bonding

In metallic bonding, bonding electrons are delocalized over a lattice of atoms. By contrast, in ionic compounds, the locations of the binding electrons and their charges are static. The freely-moving or delocalization of bonding electrons leads to classical metallic properties such as luster (surface light reflectivity), electrical and thermal conductivity, ductility, and high tensile strength.

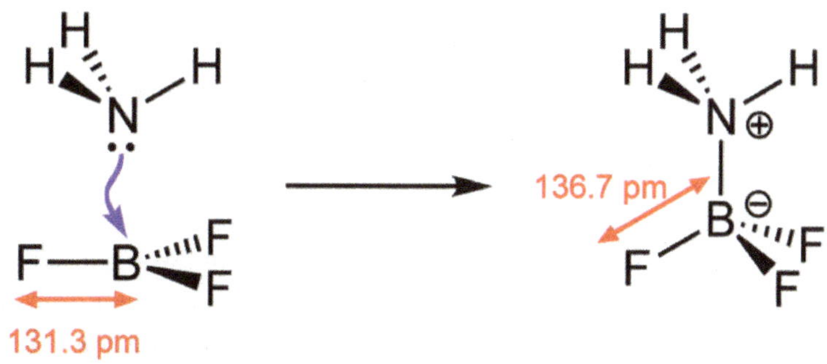

Adduct of ammonia and boron trifluoride

INTERMOLECULAR BONDING

There are four basic types of bonds that can be formed between two or more (otherwise non-associated) molecules, ions or atoms. Intermolecular forces cause molecules to be attracted or repulsed by each other. Often, these define some of the physical characteristics (such as the melting point) of a substance.

- A large difference in electronegativity between two bonded atoms will cause a permanent charge separation, or dipole, in a molecule or ion. Two or more molecules or ions with permanent dipoles can interact within dipole-dipole interactions. The bonding electrons in a molecule or ion will, on average, be closer to the more electronegative atom more frequently than the less electronegative one, giving rise to partial charges on each atom, and causing electrostatic forces between molecules or ions.

- A hydrogen bond is effectively a strong example of an interaction between two permanent dipoles. The large difference in electronegativities between hydrogen and any of fluorine, nitrogen and oxygen, coupled with their lone pairs of electrons cause strong electrostatic forces between molecules. Hydrogen bonds are responsible for the high boiling points of water and ammonia with respect to their heavier analogues.
- The London dispersion force arises due to instantaneous dipoles in neighbouring atoms. As the negative charge of the electron is not uniform around the whole atom, there is always a charge imbalance. This small charge will induce a corresponding dipole in a nearby molecule; causing an attraction between the two. The electron then moves to another part of the electron cloud and the attraction is broken.

A cation–pi interaction occurs between a pi bond and a cation.

THEORIES OF CHEMICAL BONDING

In the (unrealistic) limit of "pure" ionic bonding, electrons are perfectly localized on one of the two atoms in the bond. Such bonds can be understood by classical physics. The forces between the atoms are characterized by isotropic continuum electrostatic potentials. Their magnitude is in simple proportion to the charge difference.

Covalent bonds are better understood by valence bond theory or molecular orbital theory. The properties of the atoms involved can be understood using concepts such as oxidation number. The electron density within a bond is not assigned to individual atoms, but is instead delocalized between atoms. In valence bond theory, the two electrons on the two atoms are coupled together with the bond strength depending on the overlap between them. In molecular orbital theory, the linear combination of atomic orbitals (LCAO) helps describe the delocalized molecular orbital structures and energies based on the atomic orbitals of the atoms they came from. Unlike pure ionic bonds, covalent bonds may have directed anisotropic properties. These may have their own names, such as sigma bond and pi bond.

In the general case, atoms form bonds that are intermediate between ionic and covalent, depending on the relative electronegativity of the atoms involved. This type of bond is sometimes called polar covalent.

REFERENCES

1. Rioux, F. (2001). "The Covalent Bond in H2". The Chemical Educator. 6 (5): 288. doi:10.1007/s00897010509a.

2. Lewis, Gilbert N. (1916). "The Atom and the Molecule". Journal of the American Chemical Society. 38 (4): 772. doi:10.1021/ja02261a002. a copy
3. Laidler, K. J. (1993). The World of Physical Chemistry. Oxford University Press. p. 346. ISBN 0-19-855919-4.
4. James, H. H.; Coolidge, A. S. (1933). "The Ground State of the Hydrogen Molecule". Journal of Chemical Physics. American Institute of Physics. 1 (12): 825–835. doi:10.1063/1.1749252.
5. "Bond Lengths and Energies". Science.uwaterloo.ca. Retrieved 2013-10-15.
6. Atkins, Peter; Loretta Jones (1997). Chemistry: Molecules, Matter and Change. New York: W. H. Freeman & Co. pp. 294–295. ISBN 0-7167-3107-X.

CHAPTER 6

Chemical Reactions

Chapter Contents

- **Introduction**
- **History**
- **Equations**
- **Elementary Reactions**
- **Chemical Equilibrium**
- **Thermodynamics**
- **Kinetics**
- **Reaction Types**
- **Catalysis**
- **Biochemical Reactions**
- **Applications**
- **Monitoring**

Chemical reactions occur whenever bonds are formed or broken between molecules. Why certain atoms combine with which other atoms is a complex question which is explained exhaustively by quantum chemistry. The simple version is that atoms bond together electromagnetically based on the properties of their electron shells. There are various types of chemical bonds; in general, the more closely overlapping the electron shells are, the stronger the bond is. Chemical reactions happen when heat, radiation, and/or foreign chemicals disturb an equilibrium condition and cause the breakage and/or formation of chemical bonds, thereby giving rise to new molecular forms.

INTRODUCTION

A chemical reaction is a process that leads to the transformation of one set of chemical substances to another. Classically, chemical reactions encompass changes that only involve the positions of electrons in the forming and breaking of chemical bonds between atoms, with no change to the nuclei (no change to the elements present), and can often be described by a chemical equation. Nuclear chemistry is a sub-discipline of chemistry that involves the chemical reactions of unstable and radioactive elements where both electronic and nuclear changes may occur.

The substance (or substances) initially involved in a chemical reaction are called reactants or reagents. Chemical reactions are usually characterized by a chemical change, and they yield one or more products, which usually have properties different from the reactants. Reactions often consist of a sequence of individual sub-steps, the so-called elementary reactions, and the information on the precise course of action is part of the reaction mechanism. Chemical reactions are described with chemical equations, which symbolically present the starting materials, end products, and sometimes intermediate products and reaction conditions.

Chemical reactions happen at a characteristic reaction rate at a given temperature and chemical concentration. Typically, reaction rates increase with increasing temperature because there is more thermal energy available to reach the activation energy necessary for breaking bonds between atoms.

Reactions may proceed in the forward or reverse direction until they go to completion or reach equilibrium. Reactions that proceed in the forward direction to approach equilibrium are often described as spontaneous, requiring no input of free energy to go forward. Non-spontaneous reactions require input of free energy to go forward (examples include charging a battery by applying an external electrical power source, or photosynthesis driven by absorption of electromagnetic radiation in the form of sunlight).

Different chemical reactions are used in combinations during chemical synthesis in order to obtain a desired product. In biochemistry, a consecutive series of chemical reactions (where the product of one reaction is the reactant of the next reaction) form metabolic pathways. These reactions are often catalyzed by protein enzymes. Enzymes increase the rates of biochemical reactions, so that metabolic syntheses and decompositions impossible under ordinary conditions can occur at the temperatures and concentrations present within a cell.

A thermite reaction using iron(III) oxide. The sparks flying outwards are globules of molten iron trailing smoke in their wake.

The general concept of a chemical reaction has been extended to reactions between entities smaller than atoms, including nuclear reactions, radioactive decays, and reactions between elementary particles as described by quantum field theory.

HISTORY

Chemical reactions such as combustion in fire, fermentation and the reduction of ores to metals were known since antiquity. Initial theories of transformation of materials were developed by Greek philosophers, such as the Four-Element Theory of Empedocles stating that any substance is composed of the four basic elements – fire, water, air and earth. In the Middle Ages, chemical transformations were studied by Alchemists. They attempted, in particular, to convert lead into gold, for which purpose they used reactions of lead and lead-copper alloys with sulfur.

The production of chemical substances that do not normally occur in nature has long been tried, such as the synthesis of sulfuric and nitric acids attributed to the controversial alchemist Jābir ibn Hayyān. The process involved heating of sulfate and nitrate minerals such as copper sulfate, alum and saltpeter. In the 17th century, Johann Rudolph Glauber produced hydrochloric acid and sodium sulfate by reacting sulfuric acid and sodium chloride. With the development of the lead chamber process in 1746 and the Leblanc process, allowing large-scale production of sulfuric acid and sodium carbonate, respectively, chemical reactions became implemented into the industry. Further optimization of sulfuric acid technology resulted in the contact process in the 1880s, and the Haber process was developed in 1909–1910 for ammonia synthesis.

From the 16th century, researchers including Jan Baptist van Helmont, Robert Boyle and Isaac Newton tried to establish theories of the experimentally observed chemical transformations. The phlogiston theory was proposed in 1667 by Johann Joachim Becher. It postulated the existence of a fire-like element called "phlogiston", which was contained within combustible bodies and released during combustion. This proved to be false in 1785 by Antoine Lavoisier who found the correct explanation of the combustion as reaction with oxygen from the air.

Antoine Lavoisier developed the theory of combustion as a chemical reaction with oxygen

Joseph Louis Gay-Lussac recognized in 1808 that gases always react in a certain relationship with each other. Based on this idea and the atomic theory of John Dalton, Joseph Proust had developed the law of definite proportions, which later resulted in the concepts of stoichiometry and chemical equations.

Regarding the organic chemistry, it was long believed that compounds obtained from living organisms were too complex to be obtained synthetically. According to the concept of vitalism, organic matter was endowed with a "vital force" and distinguished from inorganic materials. This separation was ended however by the synthesis of urea from inorganic precursors by Friedrich Wöhler in 1828. Other chemists who brought major contributions to organic chemistry include Alexander William Williamson with his synthesis of ethers and Christopher Kelk Ingold, who, among many discoveries, established the mechanisms of substitution reactions.

EQUATIONS

Chemical equations are used to graphically illustrate chemical reactions. They consist of chemical or structural formulas of the reactants on the left and those of the products on the right. They are separated by an arrow (→) which indicates the direction and type of the reaction; the arrow is read as the word "yields". The tip of the arrow points in the direction in which the reaction proceeds. A double arrow (⇌) pointing in opposite directions is used for equilibrium reactions. Equations should be balanced according to the stoichiometry, the number of atoms of each species should be the same on both sides of the equation. This is achieved by scaling the number of involved molecules ({\displaystyle ABC and D in a schematic example below) by the appropriate integers a, b, c and d.

$$aA + bB \longrightarrow cC + dD$$

More elaborate reactions are represented by reaction schemes, which in addition to starting materials and products show important intermediates or transition states. Also, some relatively minor additions to the reaction can be indicated above the reaction arrow; examples of such additions are water, heat, illumination, a catalyst, etc. Similarly, some minor products can be placed below the arrow, often with a minus sign.

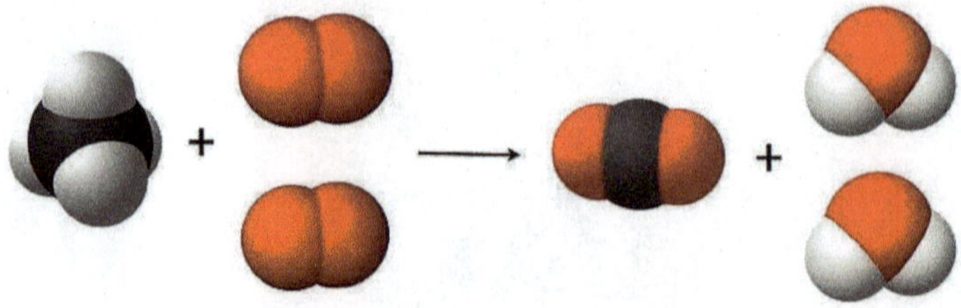

$$CH_4 + 2O_2 \longrightarrow CO_2 + 2H_2O$$

As seen from the equation $CH_4 + 2O_2 \rightarrow CO_2 + 2H_2O$, a coefficient of 2 must be placed before the oxygen gas on the reactants side and before the water on the products side in order for, as per the law of conservation of mass, the quantity of each element does not change during the reaction

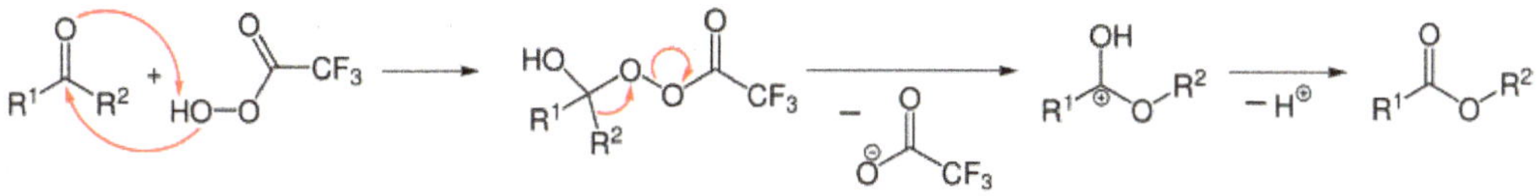

An example of organic reaction: oxidation of ketones to esters with a peroxycarboxylic acid

Retrosynthetic analysis can be applied to design a complex synthesis reaction. Here the analysis starts from the products, for example by splitting selected chemical bonds, to arrive at plausible initial reagents. A special arrow (⇒) is used in retro reactions.

ELEMENTARY REACTIONS

The elementary reaction is the smallest division into which a chemical reaction can be decomposed, it has no intermediate products. Most experimentally observed reactions are built up from many elementary reactions that occur in parallel or sequentially. The actual sequence of the individual elementary reactions is known as reaction mechanism. An elementary reaction involves a few molecules, usually one or two, because of the low probability for several molecules to meet at a certain time.

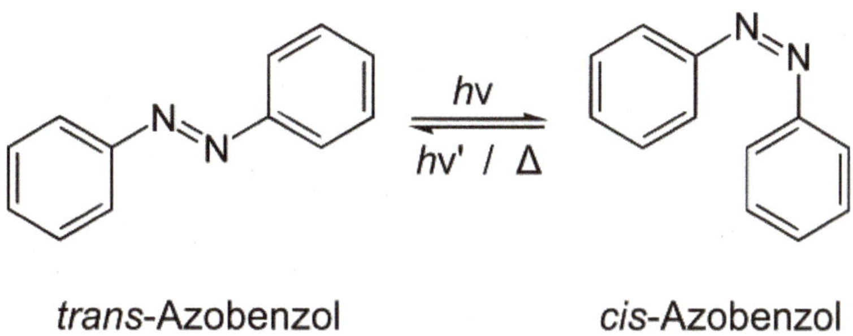

Isomerization of azobenzene, induced by light (hν) or heat (Δ)

The most important elementary reactions are unimolecular and bimolecular reactions. Only one molecule is involved in a unimolecular reaction; it is transformed by an isomerization or a dissociation into one or more other molecules. Such reactions require the addition of energy in the form of heat or light. A typical example of a unimolecular reaction is the cis–trans isomerization, in which the cis-form of a compound converts to the trans-form or vice versa.

In a typical dissociation reaction, a bond in a molecule splits (ruptures) resulting in two molecular fragments. The splitting can be homolytic or heterolytic. In the first case, the bond is divided so that each product retains an electron and becomes a neutral radical. In the second case, both electrons of the chemical bond remain with one of the products, resulting in charged ions. Dissociation plays an important role in triggering chain reactions, such as hydrogen–oxygen or polymerization reactions.

$$AB \longrightarrow A + B$$

Dissociation of a molecule AB into fragments A and B

For bimolecular reactions, two molecules collide and react with each other. Their merger is called chemical synthesis or an addition reaction.

$$A + B \longrightarrow AB$$

Another possibility is that only a portion of one molecule is transferred to the other molecule. This type of reaction occurs, for example, in redox and acid-base reactions. In redox reactions, the transferred particle is an electron, whereas in acid-base reactions it is a proton. This type of reaction is also called metathesis.

$$HA + B \longrightarrow A + HB$$

for example

$$NaCl + AgNO_3 \longrightarrow NaNO_3 + AgCl\downarrow$$

CHEMICAL EQUILIBRIUM

Most chemical reactions are reversible, that is they can and do run in both directions. The forward and reverse reactions are competing with each other and differ in reaction rates. These rates depend on the concentration and therefore change with time of the reaction: the reverse rate gradually increases and becomes equal to the rate of the forward reaction, establishing the so-called chemical equilibrium. The time to reach equilibrium depends on such parameters as temperature, pressure and the materials involved, and is determined by the minimum free energy. In equilibrium, the Gibbs free energy must be zero. The pressure dependence can be explained with the Le Chatelier's principle. For example, an increase in pressure due to decreasing volume causes the reaction to shift to the side with the fewer moles of gas.

The reaction yield stabilizes at equilibrium, but can be increased by removing the product from the reaction mixture or changed by increasing the temperature or pressure. A change in the concentrations of the reactants does not affect the equilibrium constant, but does affect the equilibrium position.

THERMODYNAMICS

Chemical reactions are determined by the laws of thermodynamics. Reactions can proceed by themselves if they are exergonic, that is if they release energy. The associated free energy of the reaction is composed of two different thermodynamic quantities, enthalpy and entropy:

$$\Delta G = \Delta H - T \cdot \Delta S.$$

G: free energy, H: enthalpy, T: temperature, S: entropy, Δ: difference(change between original and product)

Reactions can be exothermic, where ΔH is negative and energy is released. Typical examples of exothermic reactions are precipitation and crystallization, in which ordered solids are formed from disordered gaseous or liquid phases. In contrast, in endothermic reactions, heat is consumed from the environment. This can occur by increasing the entropy of the system, often through the formation of gaseous reaction products, which have high entropy. Since the entropy increases with temperature, many endothermic reactions preferably take place at high temperatures. On the contrary, many exothermic reactions such as crystallization occur at low temperatures. Changes in temperature can sometimes reverse the sign of the enthalpy of a reaction, as for the carbon monoxide reduction of molybdenum dioxide:

$$2\,CO(g) + MoO_2(s) \longrightarrow 2\,CO_2(g) + Mo(s);\ \Delta H^o = +21.86\ \text{kJ at 298 K}$$

This reaction to form carbon dioxide and molybdenum is endothermic at low temperatures, becoming less so with increasing temperature. ΔH° is zero at 1855 K, and the reaction becomes exothermic above that temperature.

Changes in temperature can also reverse the direction tendency of a reaction. For example, the water gas shift reaction

is favored by low temperatures, but its reverse is favored by high temperature. The shift in reaction direction tendency occurs at 1100 K.

Reactions can also be characterized by the internal energy which takes into account changes in the entropy, volume and chemical potential. The latter depends, among other things, on the activities of the involved substances.

$$dU = T \cdot dS - p \cdot dV + \mu \cdot dn$$

U: internal energy, S: entropy, p: pressure, μ: chemical potential, n: number of molecules, d: small change sign

KINETICS

The speed at which reactions takes place is studied by reaction kinetics. The rate depends on various parameters, such as:

- Reactant concentrations, which usually make the reaction happen at a faster rate if raised through increased collisions per unit time. Some reactions, however, have rates that are independent of reactant concentrations. These are called zero order reactions.
- Surface area available for contact between the reactants, in particular solid ones in heterogeneous systems. Larger surface areas lead to higher reaction rates.
- Pressure – increasing the pressure decreases the volume between molecules and therefore increases the frequency of collisions between the molecules.
- Activation energy, which is defined as the amount of energy required to make the reaction start and carry on spontaneously. Higher activation energy implies that the reactants need more energy to start than a reaction with a lower activation energy.

- Temperature, which hastens reactions if raised, since higher temperature increases the energy of the molecules, creating more collisions per unit time,
- The presence or absence of a catalyst. Catalysts are substances which change the pathway (mechanism) of a reaction which in turn increases the speed of a reaction by lowering the activation energy needed for the reaction to take place. A catalyst is not destroyed or changed during a reaction, so it can be used again.
- For some reactions, the presence of electromagnetic radiation, most notably ultraviolet light, is needed to promote the breaking of bonds to start the reaction. This is particularly true for reactions involving radicals.

Several theories allow calculating the reaction rates at the molecular level. This field is referred to as reaction dynamics. The rate v of a first-order reaction, which could be disintegration of a substance A, is given by:

$$v = -\frac{d[\mathrm{A}]}{dt} = k \cdot [\mathrm{A}].$$

Its integration yields:

$$[\mathrm{A}](t) = [\mathrm{A}]_0 \cdot e^{-k \cdot t}.$$

Here k is first-order rate constant having dimension 1/time, [A](t) is concentration at a time t and $[A]_0$ is the initial concentration. The rate of a first-order reaction depends only on the concentration and the properties of the involved substance, and the reaction itself can be described with the characteristic half-life. More than one time constant is needed when describing reactions of higher order. The temperature dependence of the rate constant usually follows the Arrhenius equation:

$$k = k_0 e^{-E_a/k_B T}$$

where E_a is the activation energy and kB is the Boltzmann constant. One of the simplest models of reaction rate is the collision theory. More realistic models are tailored to a specific problem and include the transition state theory, the calculation of the potential energy surface, the Marcus theory and the Rice–Ramsperger–Kassel–Marcus (RRKM) theory.

REACTION TYPES

Four Basic Types

Synthesis

In a synthesis reaction, two or more simple substances combine to form a more complex substance. These reactions are in the general form:

$$A + B \longrightarrow AB$$

Two or more reactants yielding one product is another way to identify a synthesis reaction. One example of a synthesis reaction is the combination of iron and sulfur to form iron(II) sulfide:

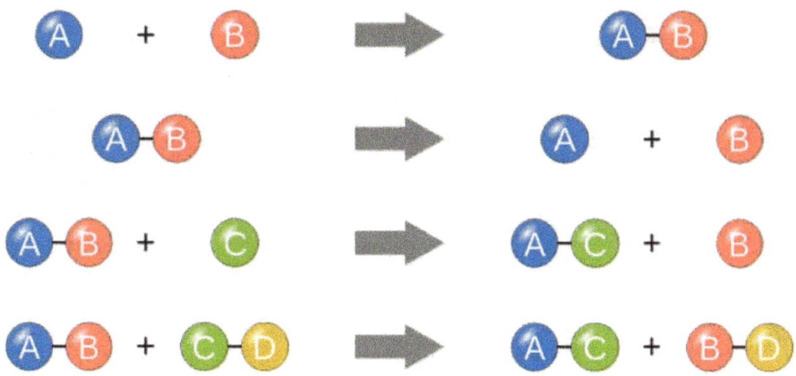

Representation of four basic chemical reactions types: synthesis, decomposition, single replacement and double replacement.

$$8\,Fe + S_8 \longrightarrow 8\,FeS$$

Another example is simple hydrogen gas combined with simple oxygen gas to produce a more complex substance, such as water.

Decomposition

A decomposition reaction is when a more complex substance breaks down into its more simple parts. It is thus the opposite of a synthesis reaction, and can be written as

$$AB \longrightarrow A + B$$

One example of a decomposition reaction is the electrolysis of water to make oxygen and hydrogen gas:

$$2\,H_2O \longrightarrow 2\,H_2 + O_2$$

Single Replacement

In a single replacement reaction, a single uncombined element replaces another in a compound; in other words, one element trades places with another element in a compound These reactions come in the general form of:

$$A + BC \longrightarrow AC + B$$

One example of a single displacement reaction is when magnesium replaces hydrogen in water to make magnesium hydroxide and hydrogen gas:

$$Mg + 2\,H_2O \longrightarrow Mg(OH)_2 + H_2 \uparrow$$

Double Replacement

In a double replacement reaction, the anions and cations of two compounds switch places and form two entirely different compounds. These reactions are in the general form:

$$AB + CD \longrightarrow AD + CB$$

For example, when barium chloride ($BaCl_2$) and magnesium sulfate ($MgSO_4$) react, the SO_4^{2-} anion switches places with the $2Cl^-$ anion, giving the compounds $BaSO_4$ and $MgCl_2$.

Another example of a double displacement reaction is the reaction of lead(II) nitrate with potassium iodide to form lead(II) iodide and potassium nitrate:

$$Pb(NO_3)_2 + 2\,KI \longrightarrow PbI_2\downarrow + 2\,KNO_3$$

Oxidation and Reduction

Redox reactions can be understood in terms of transfer of electrons from one involved species (reducing agent) to another (oxidizing agent). In this process, the former species is oxidized and the latter is reduced. Though sufficient for many purposes, these descriptions are not precisely correct. Oxidation is better defined as an increase in oxidation state, and reduction as a decrease in oxidation state. In practice, the transfer of electrons will always change the oxidation state, but there are many reactions that are classed as "redox" even though no electron transfer occurs (such as those involving covalent bonds).

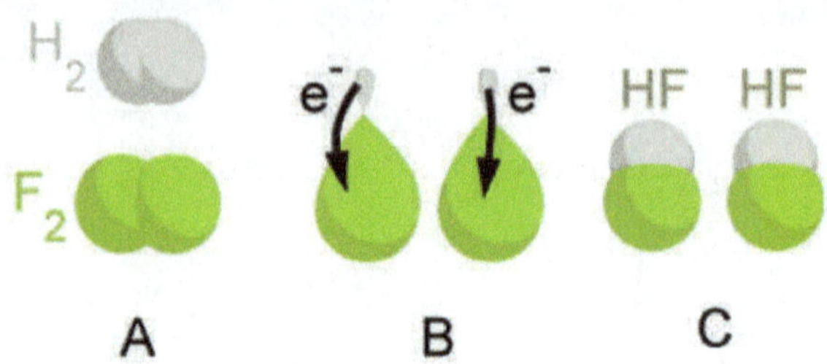

Illustration of a redox reaction

In the following redox reaction, hazardous sodium metal reacts with toxic chlorine gas to form the ionic compound sodium chloride, or common table salt:

$$2\,Na(s) + Cl_2(g) \longrightarrow 2\,NaCl(s)$$

In the reaction, sodium metal goes from an oxidation state of 0 (as it is a pure element) to +1: in other words, the sodium lost one electron and is said to have been oxidized. On the other hand, the chlorine gas goes from an oxidation of 0 (it is also a pure element) to −1: the chlorine gains one electron and is said to have been reduced. Because the chlorine is the one reduced, it is considered the electron acceptor, or in other words, induces oxidation in the sodium – thus the chlorine gas is considered the oxidizing agent. Conversely, the sodium is oxidized or is the electron donor, and thus induces reduction in the other species and is considered the reducing agent.

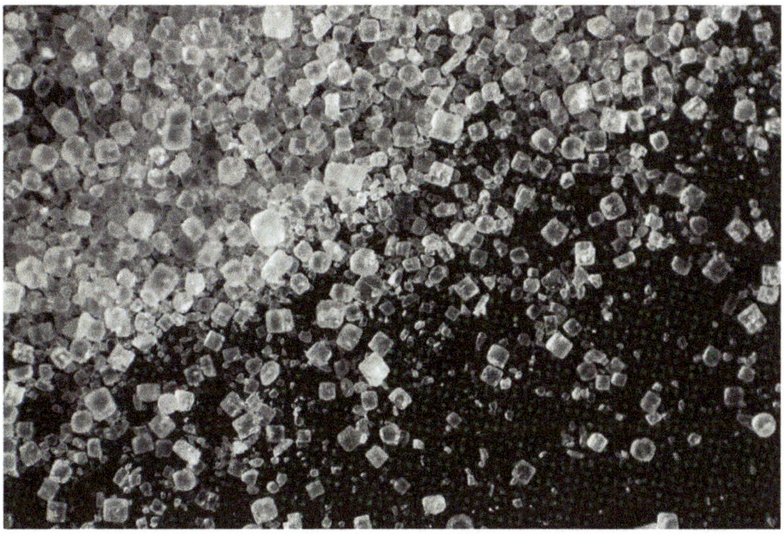

Sodium chloride is formed through the redox reaction of sodium metal and chlorine gas

Which of the involved reactants would be reducing or oxidizing agent can be predicted from the electronegativity of their elements. Elements with low electronegativity, such as most metals, easily donate electrons and oxidize – they are reducing agents. On the contrary, many ions with high oxidation numbers, such as H_2O_2, MnO_4^-, CrO_3, $Cr_2O_7^{2-}$, OsO_4 can gain one or two extra electrons and are strong oxidizing agents.

The number of electrons donated or accepted in a redox reaction can be predicted from the electron configuration of the reactant element. Elements try to reach the low-energy noble gas configuration, and therefore alkali metals and halogens will donate and accept one electron respectively. Noble gases themselves are chemically inactive.

An important class of redox reactions are the electrochemical reactions, where electrons from the power supply are used as the reducing agent. These reactions are particularly important for the production of chemical elements, such as chlorine or aluminium. The reverse process in which electrons are released in redox reactions and can be used as electrical energy is possible and used in batteries.

Complexation

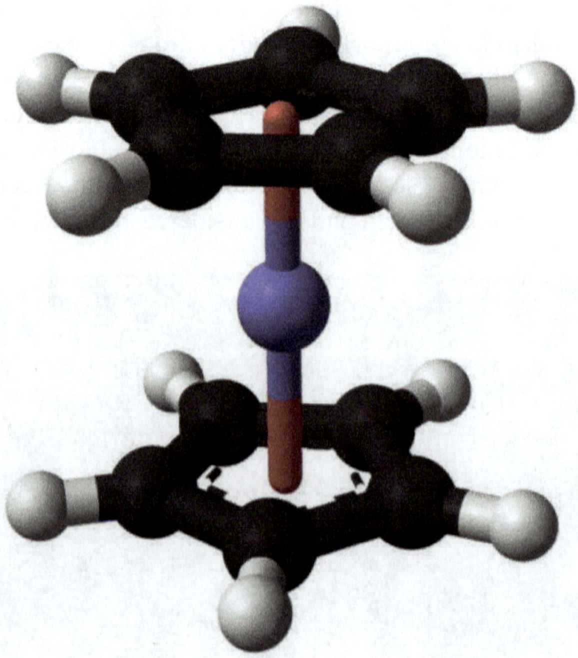

Ferrocene – an iron atom sandwiched between two C5H5 ligands

n complexation reactions, several ligands react with a metal atom to form a coordination complex. This is achieved by providing lone pairs of the ligand into empty orbitals of the metal atom and forming dipolar bonds. The ligands are Lewis bases, they can be both ions and neutral molecules, such as carbon monoxide, ammonia or water. The number of ligands that react with a central metal atom can be found using the 18-electron rule, saying that the valence shells of a transition metal will collectively accommodate 18 electrons, whereas the symmetry of the resulting complex can be predicted with the crystal field theory and ligand field theory. Complexation reactions also include ligand exchange, in which one or more ligands are replaced by another, and redox processes which change the oxidation state of the central metal atom.

Acid-Base Reactions

In the Brønsted–Lowry acid–base theory, an acid-base reaction involves a transfer of protons (H+) from one species (the acid) to another (the base). When a proton is removed from an acid, the resulting species is termed that acid's conjugate base. When the proton is accepted by a base, the resulting species is termed that base's conjugate acid. In other words, acids act as proton donors and bases act as proton acceptors according to the following equation:

$$\underbrace{HA}_{\text{acid}} + \underbrace{B}_{\text{base}} \rightleftharpoons \underbrace{A^-}_{\text{conjugated base}} + \underbrace{HB^+}_{\text{conjugated acid}}$$

The reverse reaction is possible, and thus the acid/base and conjugated base/acid are always in equilibrium. The equilibrium is determined by the acid and base dissociation constants (K_a and K_b) of the involved substances. A special case of the acid-base reaction is the neutralization where an acid and a base, taken at exactly same amounts, form a neutral salt.

- Acid-base reactions can have different definitions depending on the acid-base concept employed. Some of the most common are:
- Arrhenius definition: Acids dissociate in water releasing H3O+ ions; bases dissociate in water releasing OH− ions.
- Brønsted-Lowry definition: Acids are proton (H+) donors, bases are proton acceptors; this includes the Arrhenius definition.
- Lewis definition: Acids are electron-pair acceptors, bases are electron-pair donors; this includes the Brønsted-Lowry definition.

Precipitation

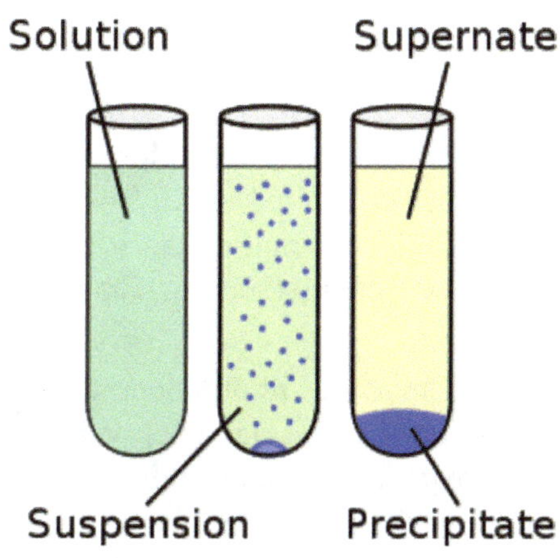

Precipitation

Precipitation is the formation of a solid in a solution or inside another solid during a chemical reaction. It usually takes place when the concentration of dissolved ions exceeds the solubility limit and forms an insoluble salt. This process can be assisted by adding a precipitating agent or by removal of the solvent. Rapid precipitation results in an amorphous or microcrystalline residue and slow process can yield single crystals. The latter can also be obtained by recrystallization from microcrystalline salts.

Solid-State Reactions

Reactions can take place between two solids. However, because of the relatively small diffusion rates in solids, the corresponding chemical reactions are very slow in comparison to liquid and gas phase reactions. They are accelerated by increasing the reaction temperature and finely dividing the reactant to increase the contacting surface area.

Reactions at the Solid|Gas Interface

Reaction can take place at the solid|gas interface, surfaces at very low pressure such as ultra-high vacuum. Via scanning tunneling microscopy, it is possible to observe reactions at the solid|gas interface in real space, if the time scale of the reaction is in the correct range. Reactions at the solid|gas interface are in some cases related to catalysis.

Photochemical Reactions

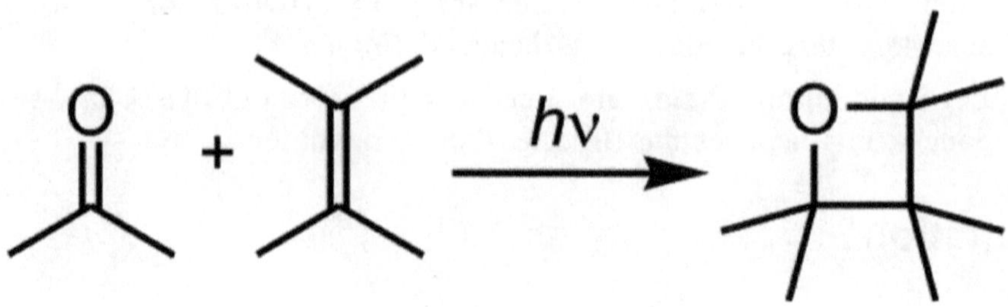

In this Paterno–Büchi reaction, a photoexcited carbonyl group is added to an unexcited olefin, yielding an oxetane.

In photochemical reactions, atoms and molecules absorb energy (photons) of the illumination light and convert into an excited state. They can then release this energy by breaking chemical bonds, thereby producing radicals. Photochemical reactions include hydrogen–oxygen reactions, radical polymerization, chain reactions and rearrangement reactions.

Many important processes involve photochemistry. The premier example is photosynthesis, in which most plants use solar energy to convert carbon dioxide and water into glucose, disposing of oxygen as a side-product. Humans rely on photochemistry for the formation of vitamin D, and vision is initiated by a photochemical reaction of rhodopsin. In fireflies, an enzyme in the abdomen catalyzes a reaction that results in bioluminescence. Many significant photochemical reactions, such as ozone formation, occur in the Earth atmosphere and constitute atmospheric chemistry.

CATALYSIS

In catalysis, the reaction does not proceed directly, but through reaction with a third substance known as catalyst. Although the catalyst takes part in the reaction, it is returned to its original state by the end of the reaction and so is not consumed. However, it can be inhibited, deactivated or destroyed by secondary processes. Catalysts can be used in a different phase (heterogeneous) or in the same phase (homogeneous) as the reactants. In heterogeneous catalysis, typical secondary processes include coking where the catalyst becomes covered by polymeric side products. Additionally, heterogeneous catalysts can dissolve into the solution in a solid–liquid system or evaporate in a solid–gas system. Catalysts can only speed up the reaction – chemicals that slow down the reaction are called inhibitors. Substances that increase the activity of catalysts are called promoters, and substances that deactivate catalysts are called catalytic poisons. With a catalyst, a reaction which is kinetically inhibited by a high activation energy can take place in circumvention of this activation energy.

Heterogeneous catalysts are usually solids, powdered in order to maximize their surface area. Of particular importance in heterogeneous catalysis are the platinum group metals and other transition metals, which are used in hydrogenations, catalytic reforming and in the synthesis of commodity chemicals such as nitric acid and ammonia. Acids are an example of a homogeneous catalyst, they increase the nucleophilicity of carbonyls, allowing a reaction that would not otherwise proceed with electrophiles. The advantage of homogeneous catalysts is the ease of mixing them with the reactants, but they may also be difficult to separate from the products. Therefore, heterogeneous catalysts are preferred in many industrial processes.

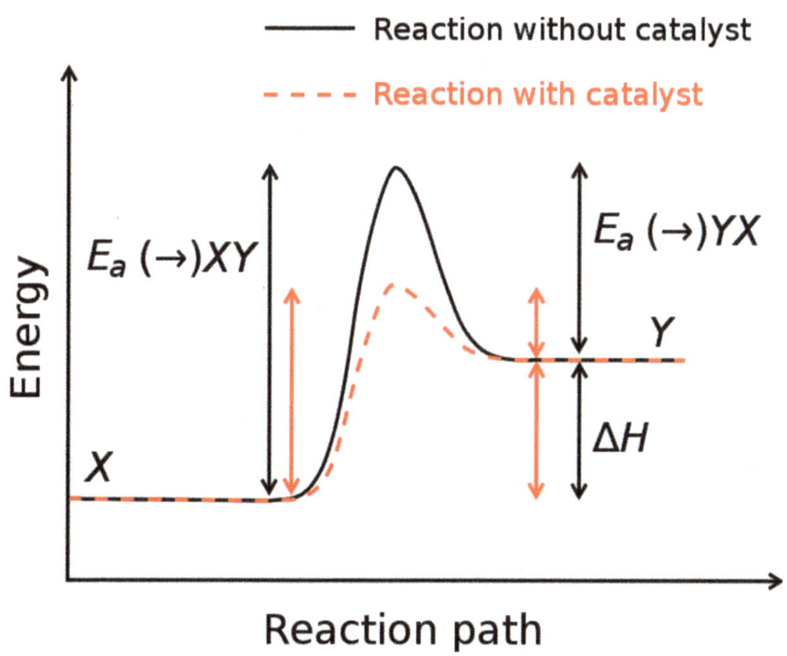

Schematic potential energy diagram showing the effect of a catalyst in an endothermic chemical reaction. The presence of a catalyst opens a different reaction pathway (in red) with a lower activation energy. The final result and the overall thermodynamics are the same.

Solid heterogeneous catalysts are plated on meshes in ceramic catalytic converters in order to maximize their surface area. This exhaust converter is from a Peugeot 106 S2 1100

Reactions in Organic Chemistry

In organic chemistry, in addition to oxidation, reduction or acid-base reactions, a number of other reactions can take place which involve covalent bonds between carbon atoms or carbon and heteroatoms (such as oxygen, nitrogen, halogens, etc.). Many specific reactions in organic chemistry are name reactions designated after their discoverers.

Substitution

In a substitution reaction, a functional group in a particular chemical compound is replaced by another group. These reactions can be distinguished by the type of substituting species into a nucleophilic, electrophilic or radical substitution.

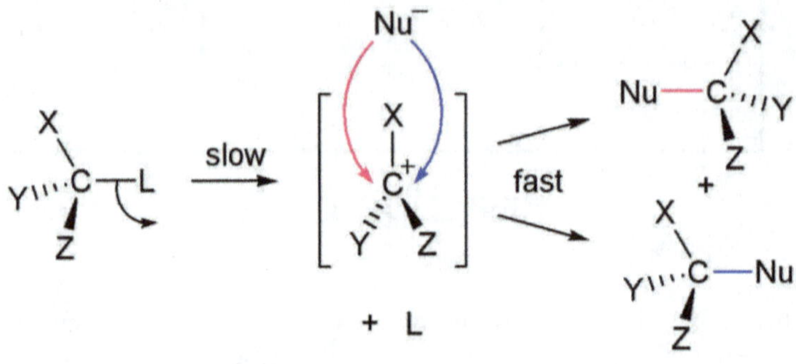

SN1 mechanism

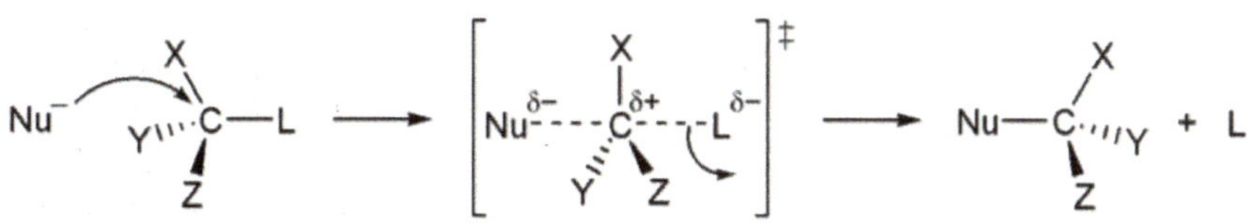

SN2 mechanism

In the first type, a nucleophile, an atom or molecule with an excess of electrons and thus a negative charge or partial charge, replaces another atom or part of the "substrate" molecule. The electron pair from the nucleophile attacks the substrate forming a new bond, while the leaving group departs with an electron pair. The nucleophile may be electrically neutral or negatively charged, whereas the substrate is typically neutral or positively charged. Examples of nucleophiles are hydroxide ion, alkoxides, amines and halides. This type of reaction is found mainly in aliphatic hydrocarbons, and rarely in aromatic hydrocarbon. The latter have high electron density and enter nucleophilic aromatic substitution only with very strong electron withdrawing groups. Nucleophilic substitution can take place by two different mechanisms, ${}^{S}N_1$ and ${}^{S}N_2$. In their names, S stands for substitution, N for nucleophilic, and the number represents the kinetic order of the reaction, unimolecular or bimolecular.

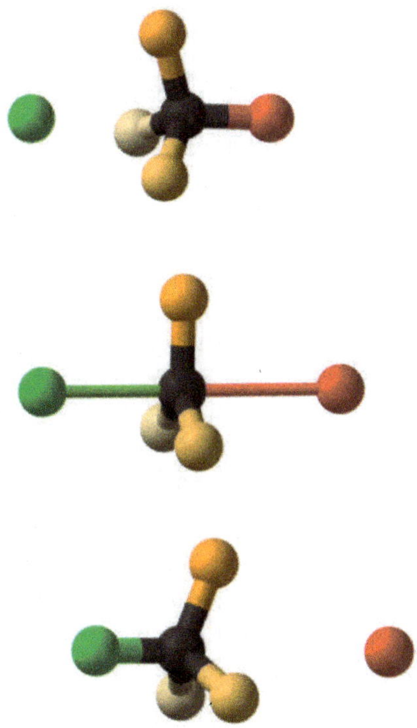

The three steps of an ${}^{S}N_2$ reaction. The nucleophile is green and the leaving group is red

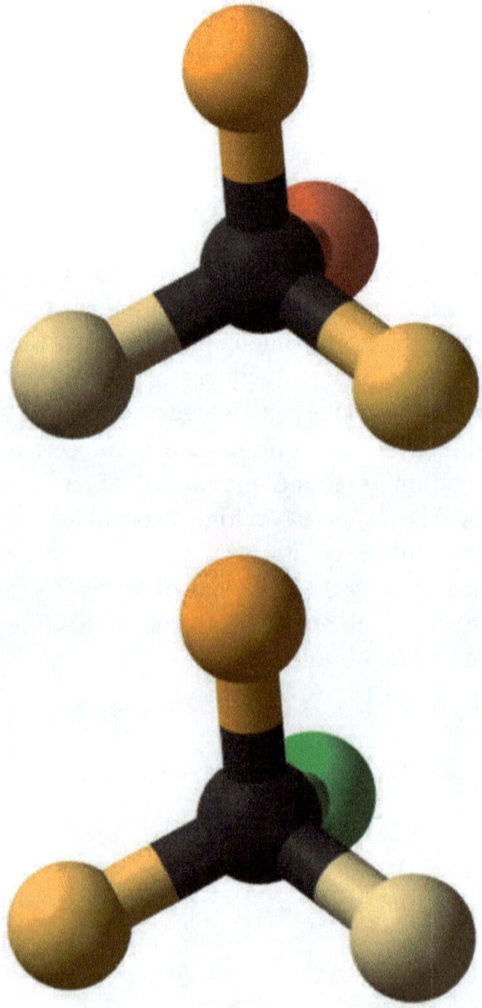

$^{S}N_2$ reaction causes stereo inversion (Walden inversion)

The $^{S}N_1$ reaction proceeds in two steps. First, the leaving group is eliminated creating a carbocation. This is followed by a rapid reaction with the nucleophile.

In the $^{S}N_2$ mechanism, the nucleophile forms a transition state with the attacked molecule, and only then the leaving group is cleaved. These two mechanisms differ in the stereochemistry of the products. $^{S}N_1$ leads to the non-stereospecific addition and does not result in a chiral center, but rather in a set of geometric isomers (cis/trans). In contrast, a reversal (Walden inversion) of the previously existing stereochemistry is observed in the $^{S}N_2$ mechanism.

Electrophilic substitution is the counterpart of the nucleophilic substitution in that the attacking atom or molecule, an electrophile, has low electron density and thus a positive charge. Typical electrophiles are the carbon atom of carbonyl groups, carbocations or sulfur or nitronium cations. This reaction takes place almost exclusively in aromatic hydrocarbons, where it is called electrophilic aromatic substitution. The electrophile attack results in the so-called σ-complex, a transition state in which the aromatic system is abolished. Then, the leaving group, usually a proton, is split off and the aromaticity is restored. An alternative to aromatic substitution is electrophilic aliphatic substitution. It is similar to the nucleophilic aliphatic substitution and also has two major types, ${}^{S}E_1$ and ${}^{S}E_2$

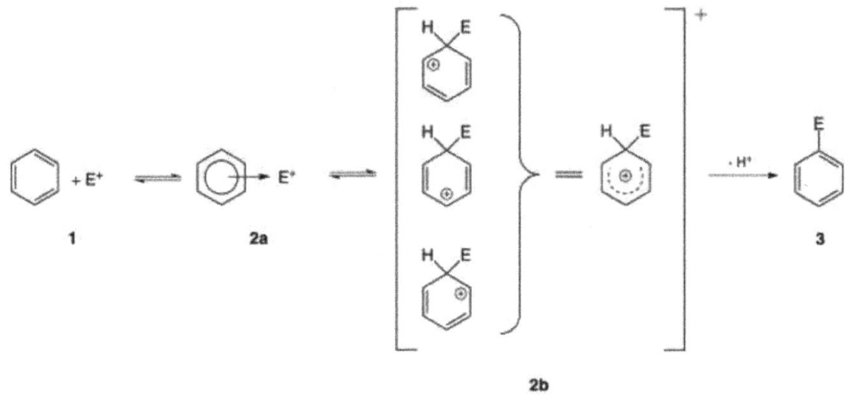

Mechanism of electrophilic aromatic substitution

In the third type of substitution reaction, radical substitution, the attacking particle is a radical. This process usually takes the form of a chain reaction, for example in the reaction of alkanes with halogens. In the first step, light or heat disintegrates the halogen-containing molecules producing the radicals. Then the reaction proceeds as an avalanche until two radicals meet and recombine.

$$X\cdot + R{-}H \longrightarrow X{-}H + R\cdot$$
$$R\cdot + X_2 \longrightarrow R{-}X + X\cdot$$

Reactions during the chain reaction of radical substitution

Addition and Elimination

The addition and its counterpart, the elimination, are reactions which change the number of substitutents on the carbon atom, and form or cleave multiple bonds. Double and triple bonds can be produced by eliminating a suitable leaving group. Similar to the nucleophilic substitution, there are several possible reaction mechanisms which are named after the respective reaction order. In the E1 mechanism, the leaving group is ejected first, forming a carbocation. The next step, formation of the double bond, takes place with elimination of a proton (deprotonation). The leaving order is reversed in the E1cb mechanism that is the proton is split off first. This mechanism requires participation of a base. Because of the similar conditions, both reactions in the E1 or E1cb elimination always compete with the ${}^{S}N_1$ substitution.

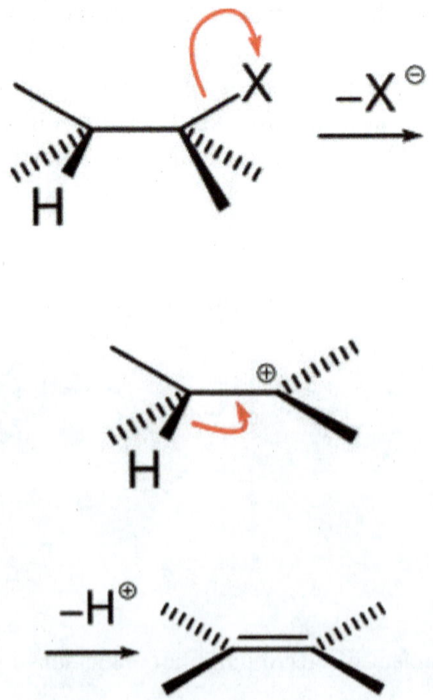

E1 elimination

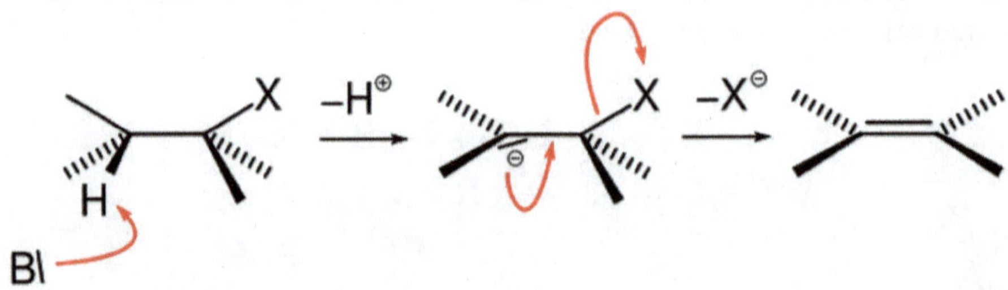

E1cb elimination

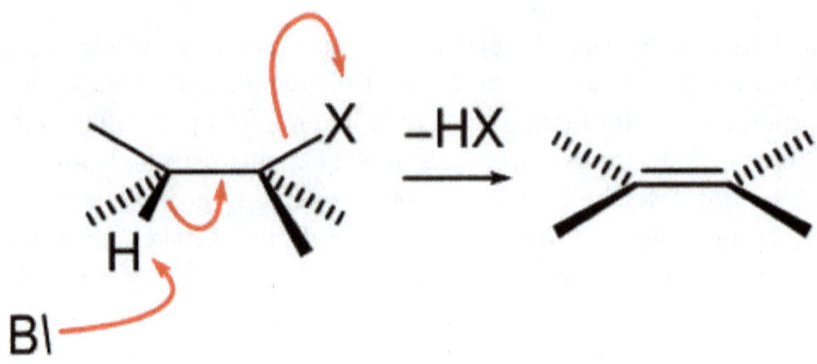

E2 elimination

The E2 mechanism also requires a base, but there the attack of the base and the elimination of the leaving group proceed simultaneously and produce no ionic intermediate. In contrast to the E1 eliminations, different stereochemical configurations are possible for the reaction product in the E2 mechanism, because the attack of the base preferentially occurs in the anti-position with respect to the leaving group. Because of the similar conditions and reagents, the E2 elimination is always in competition with the $^{S}N_2$-substitution.

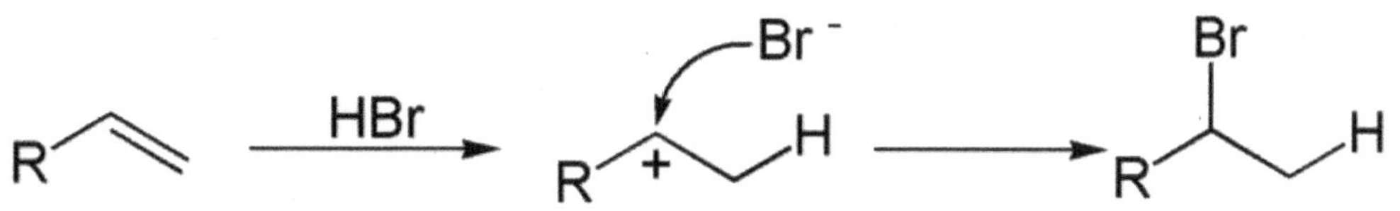

Electrophilic addition of hydrogen bromide

The counterpart of elimination is the addition where double or triple bonds are converted into single bonds. Similar to the substitution reactions, there are several types of additions distinguished by the type of the attacking particle. For example, in the electrophilic addition of hydrogen bromide, an electrophile (proton) attacks the double bond forming a carbocation, which then reacts with the nucleophile (bromine). The carbocation can be formed on either side of the double bond depending on the groups attached to its ends, and the preferred configuration can be predicted with the Markovnikov's rule. This rule states that "In the heterolytic addition of a polar molecule to an alkene or alkyne, the more electronegative (nucleophilic) atom (or part) of the polar molecule becomes attached to the carbon atom bearing the smaller number of hydrogen atoms."

If the addition of a functional group takes place at the less substituted carbon atom of the double bond, then the electrophilic substitution with acids is not possible. In this case, one has to use the hydroboration–oxidation reaction, where in the first step, the boron atom acts as electrophile and adds to the less substituted carbon atom. At the second step, the nucleophilic hydroperoxide or halogen anion attacks the boron atom.

While the addition to the electron-rich alkenes and alkynes is mainly electrophilic, the nucleophilic addition plays an important role for the carbon-heteroatom multiple bonds, and especially its most important representative, the carbonyl group. This process is often associated with an elimination, so that after the reaction the carbonyl group is present again. It is therefore called addition-elimination reaction and may occur in carboxylic acid derivatives such as chlorides, esters or anhydrides. This reaction is often catalyzed by acids or bases, where the acids increase by the electrophilicity of the carbonyl group by binding to the oxygen atom, whereas the bases enhance the nucleophilicity of the attacking nucleophile.

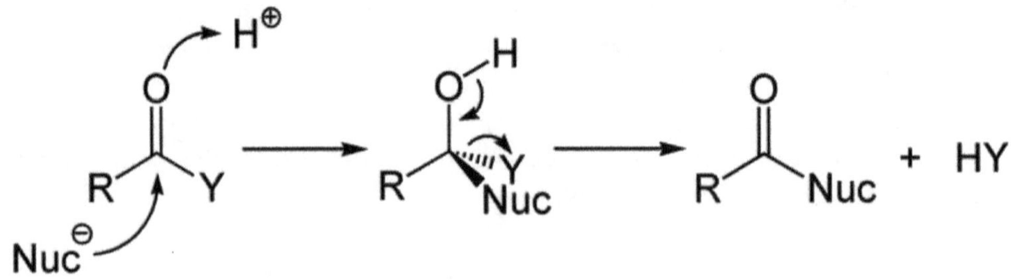

Acid-catalyzed addition-elimination mechanism

Nucleophilic addition of a carbanion or another nucleophile to the double bond of an alpha, beta unsaturated carbonyl compound can proceed via the Michael reaction, which belongs to the larger class of conjugate additions. This is one of the most useful methods for the mild formation of C–C bonds.

Some additions which cannot be executed with nucleophiles and electrophiles, can be succeeded with free radicals. As with the free-radical substitution, the radical addition proceeds as a chain reaction, and such reactions are the basis of the free-radical polymerization.

Other Organic Reaction Mechanisms

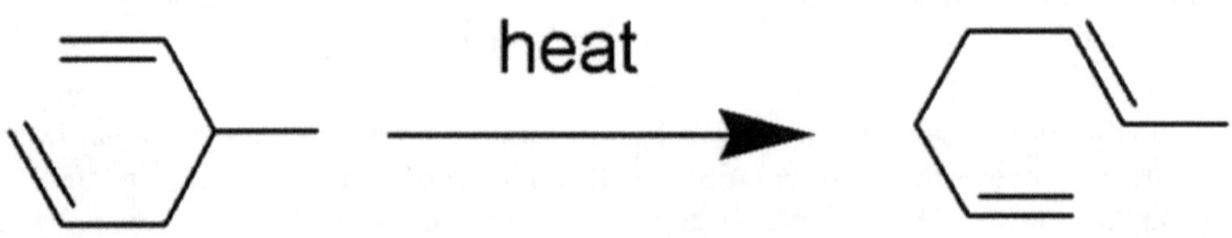

The Cope rearrangement of 3-methyl-1,5-hexadiene

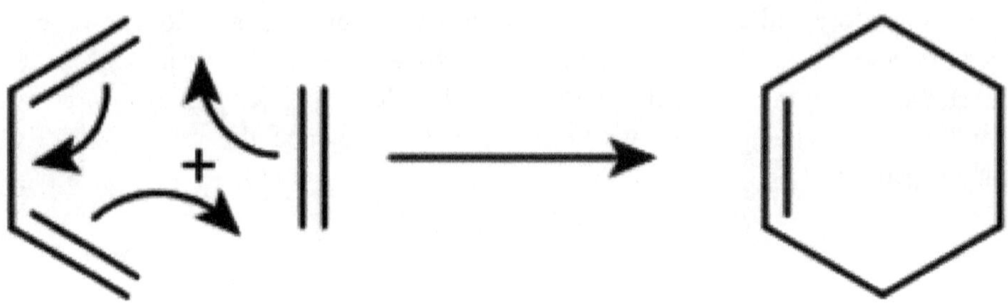

Mechanism of a Diels-Alder reaction

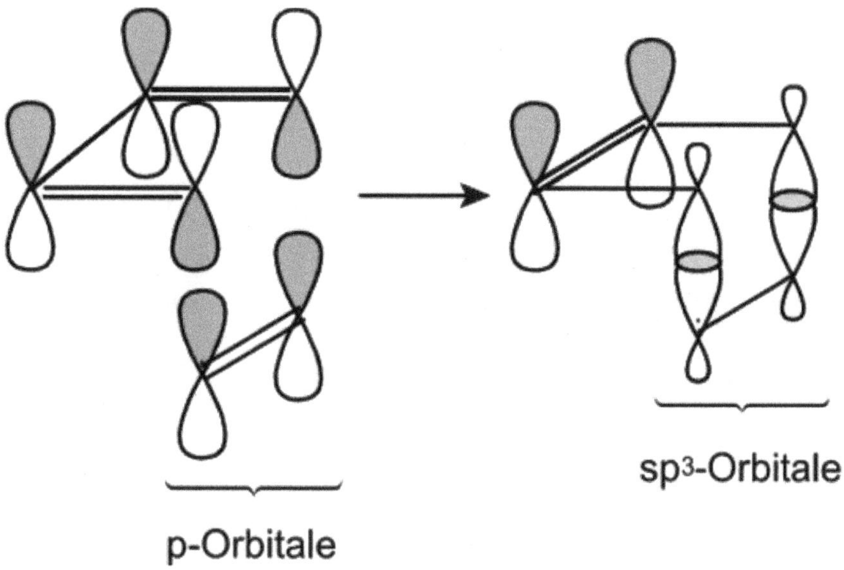

Orbital overlap in a Diels-Alder reaction

In a rearrangement reaction, the carbon skeleton of a molecule is rearranged to give a structural isomer of the original molecule. These include hydride shift reactions such as the Wagner-Meerwein rearrangement, where a hydrogen, alkyl or aryl group migrates from one carbon to a neighboring carbon. Most rearrangements are associated with the breaking and formation of new carbon-carbon bonds. Other examples are sigmatropic reaction such as the Cope rearrangement.

Cyclic rearrangements include cycloadditions and, more generally, pericyclic reactions, wherein two or more double bond-containing molecules form a cyclic molecule. An important example of cycloaddition reaction is the Diels–Alder reaction (the so-called [4+2] cycloaddition) between a conjugated diene and a substituted alkene to form a substituted cyclohexene system.

Whether a certain cycloaddition would proceed depends on the electronic orbitals of the participating species, as only orbitals with the same sign of wave function will overlap and interact constructively to form new bonds. Cycloaddition is usually assisted by light or heat. These perturbations result in different arrangement of electrons in the excited state of the involved molecules and therefore in different effects. For example, the [4+2] Diels-Alder reactions can be assisted by heat whereas the [2+2] cycloaddition is selectively induced by light. Because of the orbital character, the potential for developing stereoisomeric products upon cycloaddition is limited, as described by the Woodward–Hoffmann rules.

BIOCHEMICAL REACTIONS

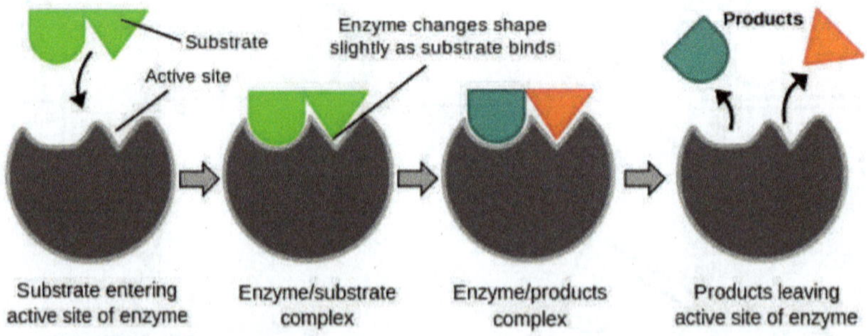

Illustration of the induced fit model of enzyme activity

Biochemical reactions are mainly controlled by enzymes. These proteins can specifically catalyze a single reaction, so that reactions can be controlled very precisely. The reaction takes place in the active site, a small part of the enzyme which is usually found in a cleft or pocket lined by amino acid residues, and the rest of the enzyme is used mainly for stabilization. The catalytic action of enzymes relies on several mechanisms including the molecular shape ("induced fit"), bond strain, proximity and orientation of molecules relative to the enzyme, proton donation or withdrawal (acid/base catalysis), electrostatic interactions and many others.

The biochemical reactions that occur in living organisms are collectively known as metabolism. Among the most important of its mechanisms is the anabolism, in which different DNA and enzyme-controlled processes result in the production of large molecules such as proteins and carbohydrates from smaller units. Bioenergetics studies the sources of energy for such reactions. An important energy source is glucose, which can be produced by plants via photosynthesis or assimilated from food. All organisms use this energy to produce adenosine triphosphate (ATP), which can then be used to energize other reactions.

APPLICATIONS

Thermite reaction proceeding in railway welding. Shortly after this, the liquid iron flows into the mould around the rail gap

Chemical reactions are central to chemical engineering where they are used for the synthesis of new compounds from natural raw materials such as petroleum and mineral ores. It is essential to make the reaction as efficient as possible, maximizing the yield and minimizing the amount of reagents, energy inputs and waste. Catalysts are especially helpful for reducing the energy required for the reaction and increasing its reaction rate.

Some specific reactions have their niche applications. For example, the thermite reaction is used to generate light and heat in pyrotechnics and welding. Although it is less controllable than the more conventional oxy-fuel welding, arc welding and flash welding, it requires much less equipment and is still used to mend rails, especially in remote areas.

MONITORING

Mechanisms of monitoring chemical reactions depend strongly on the reaction rate. Relatively slow processes can be analyzed in situ for the concentrations and identities of the individual ingredients. Important tools of real time analysis are the measurement of pH and analysis of optical absorption (color) and emission spectra. A less accessible but rather efficient method is introduction of a radioactive isotope into the reaction and monitoring how it changes over time and where it moves to; this method is often used to analyze redistribution of substances in the human body. Faster reactions are usually studied with ultrafast laser spectroscopy where utilization of femtosecond lasers allows short-lived transition states to be monitored at time scaled down to a few femtoseconds.

REFERENCES

1. IUPAC, Compendium of Chemical Terminology, 2nd ed. (the "Gold Book") (1997). Online corrected version: (2006–) "chemical reaction".
2. Weyer, J. (1973). "Neuere Interpretationsmöglichkeiten der Alchemie". Chemie in unserer Zeit. 7 (6): 177–181. doi:10.1002/ciuz.19730070604.
3. Friedman, Leonard J.; Friedman, Samantha J. (2008). The History of the Contact Sulfuric Acid Process (PDF). Boca Raton, Florida: Acid Engineering & Consulting, Inc.
4. Stranges, Anthony N. (2000). "Germany's synthetic fuel industry, 1935–1940". In Lesch, John E. The German Chemical Industry in the Twentieth Century. Kluwer Academic Publishers. p. 170. ISBN 0-7923-6487-2.
5. Brock, pp. 34–55
6. Brock, pp. 104–107
7. Myers, Richard (2009). The Basics of Chemistry. Greenwood Publishing Group. p. 55. ISBN 0313316643.
8. IUPAC, Compendium of Chemical Terminology, 2nd ed. (the "Gold Book") (1997). Online corrected version: (2006–) "chemical reaction equation".
9. Corey, E. J. (1988). "Robert Robinson Lecture. Retrosynthetic thinking?essentials and examples". Chemical Society Reviews. 17: 111. doi:10.1039/CS9881700111.

10. IUPAC, Compendium of Chemical Terminology, 2nd ed. (the "Gold Book") (1997). Online corrected version: (2006–) "elementary reaction".
11. Frenking, Gernot (2006). "Elementarreaktionen". Römpp Chemie-Lexikon. Thieme.
12. Kandori, Hideki (2006). "Retinal Binding Proteins". In Dugave, Christophe. Cis-trans Isomerization in Biochemistry. Wiley-VCH. p. 56. ISBN 3-527-31304-4.
13. Atkins, p. 114.
14. Atkins, pp. 106–108
15. Reaction Web
16. Atkins, p. 150
17. Atkins, p. 963
18. To react or not to react? Utah State Office of Education. Retrieved 4 June 2011.
19. The six types of reaction – The Cavalcade o' Chemistry. Retrieved 11 February 2016
20. Glusker, Jenny P. (1991). "Structural Aspects of Metal Liganding to Functional Groups in Proteins". In Christian B. Anfinsen. Advances in Protein Chemistry. 42. San Diego: Academic Press. p. 7. ISBN 0-12-034242-1.
21. Guo, Liang-Hong; Allen, H.; Hill, O. (1991). "Direct Electrochemistry of Proteins and Enzymes". In A. G. Sykes. Advances in Inorganic Chemistry. 36. San Diego: Academic Press. p. 359. ISBN 0-12-023636-2.
22. Wiberg, pp. 289–290
23. Wiberg, p. 409
24. Wiberg, pp. 1180–1205
25. IUPAC, Compendium of Chemical Terminology, 2nd ed. (the "Gold Book") (1997). Online corrected version: (2006–) "conjugate acid–base pair".
26. IUPAC, Compendium of Chemical Terminology, 2nd ed. (the "Gold Book") (1997). Online corrected version: (2006–) "precipitation".
27. Wingender, Jörg; Ortanderl, Stefanie (July 2009). "Ausfällung". Römpp Chemie-Lexikon. Thieme.
28. Meyer, H. Jürgen (2007). "Festkörperchemie". In Erwin Riedel. Modern Inorganic Chemistry (in German) (3rd ed.). de Gruyter. p. 171. ISBN 978-3-11-019060-1.
29. Wintterlin, J. (1997). "Atomic and Macroscopic Reaction Rates of a Surface-Catalyzed Reaction". Science. 278 (5345): 1931–4. doi:10.1126/science.278.5345.1931. PMID 9395392.
30. Waldmann, T.; Künzel, D.; Hoster, H. E.; Groß, A.; Behm, R. J. R. (2012). "Oxidation of an Organic Adlayer: A Bird's Eye View". Journal of the American Chemical Society. 134 (21): 8817–8822. doi:10.1021/ja302593v.
31. Atkins, pp. 937–950
32. Saunders, David Stanley (2002). Insect clocks (Third ed.). Amsterdam: Elsevier. p. 179. ISBN 0-444-50407-9.

33. IUPAC, Compendium of Chemical Terminology, 2nd ed. (the "Gold Book") (1997). Online corrected version: (2006–) "catalyst".
34. IUPAC, Compendium of Chemical Terminology, 2nd ed. (the "Gold Book") (1997). Online corrected version: (2006–) "inhibitor".
35. Elschenbroich, Christoph (2008). Organometallchemie (6th ed.). Wiesbaden: Vieweg+Teubner Verlag. p. 263. ISBN 978-3-8351-0167-8.
36. ^ Jump up to:a b March, Jerry (1985), Advanced Organic Chemistry: Reactions, Mechanisms, and Structure (3rd ed.), New York: Wiley, ISBN 0-471-85472-7
37. Hartshorn, S. R. (1973). Aliphatic Nucleophilic Substitution. London: Cambridge University Press. p. 1. ISBN 0-521-09801-7.
38. Bateman, Leslie C.; Church, Mervyn G.; Hughes, Edward D.; Ingold, Christopher K.; Taher, Nazeer Ahmed (1940). "188. Mechanism of substitution at a saturated carbon atom. Part XXIII. A kinetic demonstration of the unimolecular solvolysis of alkyl halides. (Section E) a general discussion". Journal of the Chemical Society: 979. doi:10.1039/JR9400000979.
39. Brückner, pp. 63–77
40. Brückner, pp. 203–206
41. Brückner, p. 16
42. Brückner, p. 192
43. Brückner, p. 183
44. Brückner, p. 172
45. Wiberg, pp. 950, 1602
46. IUPAC, Compendium of Chemical Terminology, 2nd ed. (the "Gold Book") (1997). Online corrected version: (2006–) "Markownikoff rule".
47. Brückner, p. 125
48. Latscha, Hans Peter; Kazmaier, Uli; Klein, Helmut Alfons (2008). Organische Chemie: Chemie-basiswissen II (in German). 2 (6th ed.). Springer. p. 273. ISBN 978-3-540-77106-7.
49. "Organic Reactions". 2004. doi:10.1002/0471264180. ISBN 0-471-26418-0.
50. Hunt, Ian. "Chapter 18: Enols and Enolates — The Michael Addition reaction". University of Calgary.
51. Brückner, p. 580
52. Lechner, Manfred; Gehrke, Klaus; Nordmeier, Eckhard (2003). Macromolecular Chemistry (3rd ed.). Basel: Birkhäuser. pp. 53–65. ISBN 3-7643-6952-3.
53. Fox, Marye Anne; Whitesell, James K. (2004). Organic chemistry(Third ed.). Jones & Bartlett. p. 699. ISBN 0-7637-2197-2.
54. Diels, O. .; Alder, K. . (1928). "Synthesen in der hydroaromatischen Reihe". Justus Liebig's Annalen der Chemie. 460: 98–122. doi:10.1002/jlac.19284600106.
55. Brückner, pp. 637–647

56. Woodward, R. B.; Hoffmann, R. (1965). "Stereochemistry of Electrocyclic Reactions". Journal of the American Chemical Society. 87 (2): 395–397. doi:10.1021/ja01080a054.
57. Karlson, Peter; Doenecke, Detlef; Koolman, Jan; Fuchs, Georg; Gerok, Wolfgang (2005). Karlson Biochemistry and Pathobiochemistry (in German) (16th ed.). Thieme. pp. 55–56. ISBN 978-3-13-357815-8.
58. IUPAC, Compendium of Chemical Terminology, 2nd ed. (the "Gold Book") (1997). Online corrected version: (2006–) "anabolism".
59. Emig, Gerhard; Klemm, Elias (2005). Technical Chemistry (in German) (5th ed.). Springer. pp. 33–34. ISBN 978-3-540-23452-4.
60. Trost, B. (1991). "The atom economy--a search for synthetic efficiency". Science. 254 (5037): 1471–7. doi:10.1126/science.1962206. PMID 1962206.
61. Weismantel, Guy E (1999). John J. McKetta, ed. Encyclopedia of Chemical Processing and Design. 67. CRC Press. p. 109. ISBN 0-8247-2618-9.
62. Atkins, p. 987

CHAPTER 7

Chemical Equilibrium

Chapter Contents

- **Introduction**
- **Determination Of Equilibrium Constants**
- **Equilibrium Constant**
- **Henderson–Hasselbalch Equation**
- **Michaelis–Menten Kinetics**

Chemical equilibrium refers to a stable ratio between reactants and products in a reversible chemical reaction. In a reversible reaction, reactants do not completely transform into products; rather, they will slowly stop reacting as chemical equilibrium is reached. The rate of a reaction is affected by many factors, including temperature, phase of matter and presence of a catalyst. Many reactions require an initial input of energy to begin reacting.

INTRODUCTION

In a chemical reaction, chemical equilibrium is the state in which both reactants and products are present in concentrations which have no further tendency to change with time. Usually, this state results when the forward reaction proceeds at the same rate as the reverse reaction. The reaction rates of the forward and backward reactions are generally not zero, but equal. Thus, there are no net changes in the concentrations of the reactant(s) and product(s). Such a state is known as dynamic equilibrium.

Historical Introduction

The concept of chemical equilibrium was developed after Berthollet (1803) found that some chemical reactions are reversible. For any reaction mixture to exist at equilibrium, the rates of the forward and backward (reverse) reactions are equal. In the following chemical equation with arrows pointing both ways to indicate equilibrium, A and B are reactant chemical species, S and T are product species, and α, β, σ, and τ are the stoichiometric coefficients of the respective reactants and products:

$$\alpha A + \beta B \rightleftharpoons \sigma S + \tau T$$

The equilibrium concentration position of a reaction is said to lie "far to the right" if, at equilibrium, nearly all the reactants are consumed. Conversely the equilibrium position is said to be "far to the left" if hardly any product is formed from the reactants.

Guldberg and Waage (1865), building on Berthollet's ideas, proposed the law of mass action:

$$\text{forward reaction rate} = k_+ A^\alpha B^\beta$$
$$\text{backward reaction rate} = k_- S^\sigma T^\tau$$

where A, B, S and T are active masses and k_+ and k_- are rate constants. Since at equilibrium forward and backward rates are equal:

$$k_+ \{A\}^\alpha \{B\}^\beta = k_- \{S\}^\sigma \{T\}^\tau$$

and the ratio of the rate constants is also a constant, now known as an equilibrium constant.

$$K_c = \frac{k_+}{k_-} = \frac{\{S\}^\sigma \{T\}^\tau}{\{A\}^\alpha \{B\}^\beta}$$

By convention the products form the numerator. However, the law of mass action is valid only for concerted one-step reactions that proceed through a single transition state and is not valid in general because rate equations do not, in general, follow the stoichiometry of the reaction as Guldberg and Waage had proposed (see, for example, nucleophilic aliphatic substitution by $^{S}N_1$ or reaction of hydrogen and bromine to form hydrogen bromide). Equality of forward and backward reaction rates, however, is a necessary condition for chemical equilibrium, though it is not sufficient to explain why equilibrium occurs.

Despite the failure of this derivation, the equilibrium constant for a reaction is indeed a constant, independent of the activities of the various species involved, though it does depend on temperature as observed by the van 't Hoff equation. Adding a catalyst will affect both the forward reaction and the reverse reaction in the same way and will not have an effect on the equilibrium constant. The catalyst will speed up both reactions thereby increasing the speed at which equilibrium is reached.

Although the macroscopic equilibrium concentrations are constant in time, reactions do occur at the molecular level. For example, in the case of acetic acid dissolved in water and forming acetate and hydronium ions,

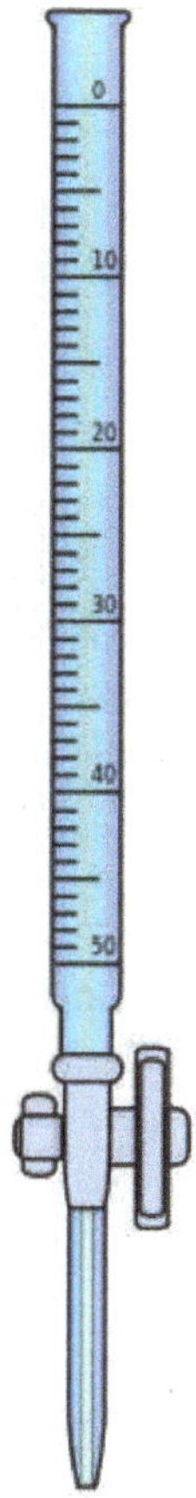

Burette, a common laboratory apparatus for carrying out titration, an important experimental technique in equilibrium and analytical chemistry.

$$CH_3CO_2H + H_2O \rightleftharpoons CH_3CO_2^- + H_3O^+$$

a proton may hop from one molecule of acetic acid on to a water molecule and then on to an acetate anion to form another molecule of acetic acid and leaving the number of acetic acid molecules unchanged. This is an example of dynamic equilibrium. Equilibria, like the rest of thermodynamics, are statistical phenomena, averages of microscopic behavior.

Le Châtelier's principle (1884) gives an idea of the behavior of an equilibrium system when changes to its reaction conditions occur. If a dynamic equilibrium is disturbed by changing the conditions, the position of equilibrium moves to partially reverse the change. For example, adding more S from the outside will cause an excess of products, and the system will try to counteract this by increasing the reverse reaction and pushing the equilibrium point backward (though the equilibrium constant will stay the same).

If mineral acid is added to the acetic acid mixture, increasing the concentration of hydronium ion, the amount of dissociation must decrease as the reaction is driven to the left in accordance with this principle. This can also be deduced from the equilibrium constant expression for the reaction:

$$K = \frac{\{CH_3CO_2^-\}\{H_3O^+\}}{\{CH_3CO_2H\}}$$

If $\{H_3O^+\}$ increases $\{CH_3CO_2H\}$ must increase and $CH_3CO_2^-$ must decrease. The H_2O is left out, as it is the solvent and its concentration remains high and nearly constant.

A quantitative version is given by the reaction quotient.

J. W. Gibbs suggested in 1873 that equilibrium is attained when the Gibbs free energy of the system is at its minimum value (assuming the reaction is carried out at constant temperature and pressure). What this means is that the derivative of the Gibbs energy with respect to reaction coordinate (a measure of the extent of reaction that has occurred, ranging from zero for all reactants to a maximum for all products) vanishes, signalling a stationary point. This derivative is called the reaction Gibbs energy (or energy change) and corresponds to the difference between the chemical potentials of reactants and products at the composition of the reaction mixture. This criterion is both necessary and sufficient. If a mixture is not at equilibrium, the liberation of the excess Gibbs energy (or Helmholtz energy at constant volume reactions) is the "driving force" for the composition of the mixture to change until equilibrium is reached. The equilibrium constant can be related to the standard Gibbs free energy change for the reaction by the equation

$$\Delta_r G^\ominus = -RT \ln K_{eq}$$

where R is the universal gas constant and T the temperature.

When the reactants are dissolved in a medium of high ionic strength the quotient of activity coefficients may be taken to be constant. In that case the concentration quotient, K_c,

$$K_c = \frac{[S]^\sigma [T]^\tau}{[A]^\alpha [B]^\beta}$$

where [A] is the concentration of A, etc., is independent of the analytical concentration of the reactants. For this reason, equilibrium constants for solutions are usually determined in media of high ionic strength. K_c varies with ionic strength, temperature and pressure (or volume). Likewise K_p for gases depends on partial pressure. These constants are easier to measure and encountered in high-school chemistry courses.

Thermodynamics

At constant temperature and pressure, one must consider the Gibbs free energy, G, while at constant temperature and volume, one must consider the Helmholtz free energy: A, for the reaction; and at constant internal energy and volume, one must consider the entropy for the reaction: S.

The constant volume case is important in geochemistry and atmospheric chemistry where pressure variations are significant. Note that, if reactants and products were in standard state (completely pure), then there would be no reversibility and no equilibrium. Indeed, they would necessarily occupy disjoint volumes of space. The mixing of the products and reactants contributes a large entropy (known as entropy of mixing) to states containing equal mixture of products and reactants. The standard Gibbs energy change, together with the Gibbs energy of mixing, determine the equilibrium state.

In this article only the constant pressure case is considered. The relation between the Gibbs free energy and the equilibrium constant can be found by considering chemical potentials.

At constant temperature and pressure, the Gibbs free energy, G, for the reaction depends only on the extent of reaction: ξ (Greek letter xi), and can only decrease according to the second law of thermodynamics. It means that the derivative of G with ξ must be negative if the reaction happens; at the equilibrium the derivative being equal to zero.

$$\left(\frac{dG}{d\xi}\right)_{T,p} = 0: \quad \text{equilibrium}$$

In order to meet the thermodynamic condition for equilibrium, the Gibbs energy must be stationary, meaning that the derivative of G with respect to the extent of reaction: ξ, must be zero. It can be shown that in this case, the sum of chemical potentials of the products is equal to the sum of those corresponding to the reactants. Therefore, the sum of the Gibbs energies of the reactants must be the equal to the sum of the Gibbs energies of the products.

$$\alpha\mu_{\mathrm{A}} + \beta\mu_{\mathrm{B}} = \sigma\mu_{\mathrm{S}} + \tau\mu_{\mathrm{T}}$$

where μ is in this case a partial molar Gibbs energy, a chemical potential. The chemical potential of a reagent A is a function of the activity, {A} of that reagent.

$$\mu_{\mathrm{A}} = \mu_A^{\ominus} + RT\ln\{\mathrm{A}\}$$

(where $\mu_{\mathrm{A}}^{\ominus}$ is the standard chemical potential).

The definition of the Gibbs energy equation interacts with the fundamental thermodynamic relation to produce

$$dG = Vdp - SdT + \sum_{i=1}^{k} \mu_i dN_i.$$

Inserting dNi = vi dξ into the above equation gives a Stoichiometric coefficient (Vi) and a differential that denotes the reaction occurring once (dξ). At constant pressure and temperature the above equations can be written as

$$\left(\frac{dG}{d\xi}\right)_{T,p} = \sum_{i=1}^{k} \mu_i \nu_i = \Delta_{\mathrm{r}} G_{T,p}$$

which is the "Gibbs free energy change for the reaction.

This results in:

$$\Delta_{\mathrm{r}} G_{T,p} = \sigma\mu_{\mathrm{S}} + \tau\mu_{\mathrm{T}} - \alpha\mu_{\mathrm{A}} - \beta\mu_{\mathrm{B}}.$$

By substituting the chemical potentials:

$$\Delta_r G_{T,p} = (\sigma\mu_S^\ominus + \tau\mu_T^\ominus) - (\alpha\mu_A^\ominus + \beta\mu_B^\ominus) + (\sigma RT\ln\{S\} + \tau RT\ln\{T\}) - (\alpha RT\ln\{A\} + \beta RT\ln\{B\}),$$

the relationship becomes:

$$\Delta_r G_{T,p} = \sum_{i=1}^{k} \mu_i^\ominus \nu_i + RT\ln\frac{\{S\}^\sigma\{T\}^\tau}{\{A\}^\alpha\{B\}^\beta}$$

$$\sum_{i=1}^{k} \mu_i^\ominus \nu_i = \Delta_r G^\ominus;$$

which is the standard Gibbs energy change for the reaction that can be calculated using thermodynamical tables. The reaction quotient is defined as:

$$Q_r = \frac{\{S\}^\sigma\{T\}^\tau}{\{A\}^\alpha\{B\}^\beta}$$

Therefore,

$$\left(\frac{dG}{d\xi}\right)_{T,p} = \Delta_r G_{T,p} = \Delta_r G^\ominus + RT\ln Q_r$$

At equilibrium:

$$\left(\frac{dG}{d\xi}\right)_{T,p} = \Delta_r G_{T,p} = 0$$

leading to:

$$0 = \Delta_r G^\ominus + RT\ln K_{eq}$$

and

$$\Delta_r G^\ominus = -RT\ln K_{eq}$$

Obtaining the value of the standard Gibbs energy change, allows the calculation of the equilibrium constant.

Addition of Reactants or Products

For a reactional system at equilibrium: $Q_r = K_{eq}$; $\xi = \xi_{eq}$.

If are modified activities of constituents, the value of the reaction quotient changes and becomes different from the equilibrium constant: $Q_r \neq K_{eq}$

$$\left(\frac{dG}{d\xi}\right)_{T,p} = \Delta_r G^\ominus + RT \ln Q_r$$

and

$$\Delta_r G^\ominus = -RT \ln K_{eq}$$

then

$$\left(\frac{dG}{d\xi}\right)_{T,p} = RT \ln\left(\frac{Q_r}{K_{eq}}\right)$$

- If activity of a reagent i increases

$$Q_r = \frac{\prod (a_j)^{\nu_j}}{\prod (a_i)^{\nu_i}}$$, the reaction quotient decreases.

then

$$Q_r < K_{eq} \quad \text{and} \quad \left(\frac{dG}{d\xi}\right)_{T,p} < 0$$

The reaction will shift to the right (i.e. in the forward direction, and thus more products will form).

- If activity of a product j increases

then

$$Q_r > K_{eq} \quad \text{and} \quad \left(\frac{dG}{d\xi}\right)_{T,p} > 0$$

The reaction will shift to the left (i.e. in the reverse direction, and thus less products will form).

Note that activities and equilibrium constants are dimensionless numbers.

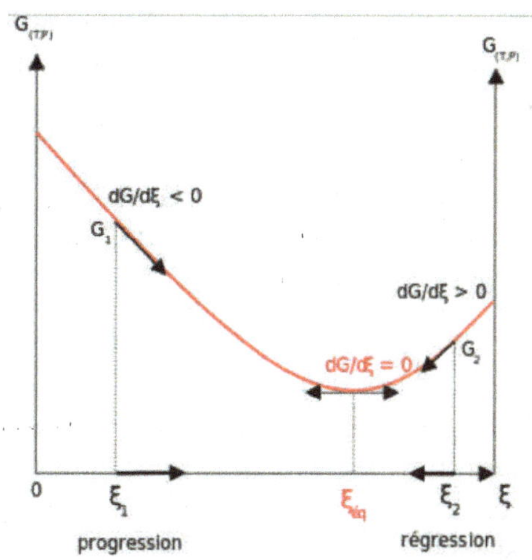

Diagramme G = f (ξ)(T,p)

Treatment of Activity

The expression for the equilibrium constant can be rewritten as the product of a concentration quotient, K_c and an activity coefficient quotient, Γ.

$$K = \frac{[S]^\sigma [T]^\tau \ldots}{[A]^\alpha [B]^\beta \ldots} \times \frac{\gamma_S{}^\sigma \gamma_T{}^\tau \ldots}{\gamma_A{}^\alpha \gamma_B{}^\beta \ldots} = K_c \Gamma$$

[A] is the concentration of reagent A, etc. It is possible in principle to obtain values of the activity coefficients, γ. For solutions, equations such as the Debye–Hückel equation or extensions such as Davies equation Specific ion interaction theory or Pitzer equations may be used.Software (below). However this is not always possible. It is common practice to assume that Γ is a constant, and to use the concentration quotient in place of the thermodynamic equilibrium constant. It is also general practice to use the term equilibrium constant instead of the more accurate concentration quotient. This practice will be followed here.

For reactions in the gas phase partial pressure is used in place of concentration and fugacity coefficient in place of activity coefficient. In the real world, for example, when making ammonia in industry, fugacity coefficients must be taken into account. Fugacity, f, is the product of partial pressure and fugacity coefficient. The chemical potential of a species in the gas phase is given by

$$\mu = \mu^\ominus + RT \ln\left(\frac{f}{\text{bar}}\right) = \mu^\ominus + RT \ln\left(\frac{p}{\text{bar}}\right) + RT \ln \gamma$$

so the general expression defining an equilibrium constant is valid for both solution and gas phases.

Concentration Quotients

In aqueous solution, equilibrium constants are usually determined in the presence of an "inert" electrolyte such as sodium nitrate $NaNO_3$ or potassium perchlorate $KClO_4$. The ionic strength of a solution is given by

$$I = \frac{1}{2} \sum_{i=1}^{N} c_i z_i^2$$

where c_i and z_i stand for the concentration and ionic charge of ion type i, and the sum is taken over all the N types of charged species in solution. When the concentration of dissolved salt is much higher than the analytical concentrations of the reagents, the ions originating from the dissolved salt determine the ionic strength, and the ionic strength is effectively constant. Since activity coefficients depend on ionic strength the activity coefficients of the species are effectively independent of concentration. Thus, the assumption that Γ is constant is justified. The concentration quotient is a simple multiple of the equilibrium constant.

$$K_c = \frac{K}{\Gamma}$$

However, K_c will vary with ionic strength. If it is measured at a series of different ionic strengths the value can be extrapolated to zero ionic strength. The concentration quotient obtained in this manner is known, paradoxically, as a thermodynamic equilibrium constant.

To use a published value of an equilibrium constant in conditions of ionic strength different from the conditions used in its determination, the value should be adjusted.

Metastable Mixtures

A mixture may be appear to have no tendency to change, though it is not at equilibrium. For example, a mixture of SO_2 and O_2 is metastable as there is a kinetic barrier to formation of the product, SO_3.

$$2\,SO_2 + O_2 \rightleftharpoons 2\,SO_3$$

The barrier can be overcome when a catalyst is also present in the mixture as in the contact process, but the catalyst does not affect the equilibrium concentrations.

Likewise, the formation of bicarbonate from carbon dioxide and water is very slow under normal conditions

$$CO_2 + 2\,H_2O \rightleftharpoons HCO_3^- + H_3O^+$$

but almost instantaneous in the presence of the catalytic enzyme carbonic anhydrase.

Pure Substances

When pure substances (liquids or solids) are involved in equilibria their activities do not appear in the equilibrium constant because their numerical values are considered one.

Applying the general formula for an equilibrium constant to the specific case of a dilute solution of acetic acid in water one obtains

$$CH_3CO_2H + H_2O \rightleftharpoons CH_3CO_2^- + H_3O^+$$

$$K_c = \frac{[CH_3CO_2^-][H_3O^+]}{[CH_3CO_2H][H_2O]}$$

For all but very concentrated solutions, the water can be considered a "pure" liquid, and therefore it has an activity of one. The equilibrium constant expression is therefore usually written as

$$K = \frac{[CH_3CO_2^-][H_3O^+]}{[CH_3CO_2H]} = K_c.$$

A particular case is the self-ionization of water itself

$$2\,H_2O \rightleftharpoons H_3O^+ + OH^-$$

Because water is the solvent, and has an activity of one, the self-ionization constant of water is defined as

$$K_w = [H^+][OH^-]$$

It is perfectly legitimate to write $[H^+]$ for the hydronium ion concentration, since the state of solvation of the proton is constant (in dilute solutions) and so does not affect the equilibrium concentrations. K_w varies with variation in ionic strength and/or temperature. The concentrations of H^+ and OH^- are not independent quantities. Most commonly $[OH^-]$ is replaced by $Kw[H^+]^-1$ in equilibrium constant expressions which would otherwise include hydroxide ion.

Solids also do not appear in the equilibrium constant expression, if they are considered to be pure and thus their activities taken to be one. An example is the Boudouard reaction:

$$2\,CO \rightleftharpoons CO_2 + C$$

for which the equation (without solid carbon) is written as:

$$K_c = \frac{[CO_2]}{[CO]^2}$$

Multiple Equilibria

Consider the case of a dibasic acid H_2A. When dissolved in water, the mixture will contain H_2A, H^{A-} and A^{2-}. This equilibrium can be split into two steps in each of which one proton is liberated.

$$H_2A \rightleftharpoons HA^- + H^+ : \quad K_1 = \frac{[HA-][H^+]}{[H_2A]}$$

$$HA- \rightleftharpoons A^{2-} + H^+ : \quad K_2 = \frac{[A^{2-}][H^+]}{[HA-]}$$

K_1 and K_2 are examples of stepwise equilibrium constants. The overall equilibrium constant, β_D, is product of the stepwise constants.

$$H_2A \rightleftharpoons A^{2-} + 2H^+ : \quad \beta_D = \frac{[A^{2-}][H^+]^2}{[H_2A]} = K_1K_2$$

Note that these constants are dissociation constants because the products on the right hand side of the equilibrium expression are dissociation products. In many systems, it is preferable to use association constants.

$$A^{2-} + H^+ \rightleftharpoons HA^- : \quad \beta_1 = \frac{[HA^-]}{[A^{2-}][H^+]}$$

$$A^{2-} + 2H^+ \rightleftharpoons H_2A : \quad \beta_2 = \frac{[H_2A]}{[A^{2-}][H^+]^2}$$

β_1 and β_2 are examples of association constants. Clearly $\beta_1 = \frac{1}{K_2}$ and $\beta_2 = \frac{1}{\beta_D}$; $\log \beta_1 = pK_2$ and $\log \beta_2 = pK_2 + pK_1$ For multiple equilibrium systems, also see: theory of Response reactions.

Effect of Temperature

The effect of changing temperature on an equilibrium constant is given by the van 't Hoff equation

$$\frac{d \ln K}{dT} = \frac{\Delta H_m^\ominus}{RT^2}$$

Thus, for exothermic reactions (ΔH is negative), K decreases with an increase in temperature, but, for endothermic reactions, (ΔH is positive) K increases with an increase temperature. An alternative formulation is

$$\frac{d \ln K}{d(T^{-1})} = -\frac{\Delta H_m^\ominus}{R}$$

At first sight this appears to offer a means of obtaining the standard molar enthalpy of the reaction by studying the variation of K with temperature. In practice, however, the method is unreliable because error propagation almost always gives very large errors on the values calculated in this way.

Types of Equilibrium

- In the gas phase: rocket engines
- The industrial synthesis such as ammonia in the Haber–Bosch process (depicted right) takes place through a succession of equilibrium steps including adsorption processes
- Atmospheric chemistry
- Seawater and other natural waters: chemical oceanography
- Distribution between two phases
- log D distribution coefficient: important for pharmaceuticals where lipophilicity is a significant property of a drug
- Liquid–liquid extraction, Ion exchange, Chromatography
- Solubility product
- Uptake and release of oxygen by haemoglobin in blood

- Acid–base equilibria: acid dissociation constant, hydrolysis, buffer solutions, indicators, acid–base homeostasis
- Metal–ligand complexation: sequestering agents, chelation therapy, MRI contrast reagents, Schlenk equilibrium
- Adduct formation: host–guest chemistry, supramolecular chemistry, molecular recognition, dinitrogen tetroxide
- In certain oscillating reactions, the approach to equilibrium is not asymptotically but in the form of a damped oscillation .
- The related Nernst equation in electrochemistry gives the difference in electrode potential as a function of redox concentrations.
- When molecules on each side of the equilibrium are able to further react irreversibly in secondary reactions, the final product ratio is determined according to the Curtin–Hammett principle.

In these applications, terms such as stability constant, formation constant, binding constant, affinity constant, association/dissociation constant are used. In biochemistry, it is common to give units for binding constants, which serve to define the concentration units used when the constant's value was determined.

Composition of a Mixture

When the only equilibrium is that of the formation of a 1:1 adduct as the composition of a mixture, there are any number of ways that the composition of a mixture can be calculated. For example, see ICE table for a traditional method of calculating the pH of a solution of a weak acid.

There are three approaches to the general calculation of the composition of a mixture at equilibrium.

- The most basic approach is to manipulate the various equilibrium constants until the desired concentrations are expressed in terms of measured equilibrium constants (equivalent to measuring chemical potentials) and initial conditions.
- Minimize the Gibbs energy of the system.
- Satisfy the equation of mass balance. The equations of mass balance are simply statements that demonstrate that the total concentration of each reactant must be constant by the law of conservation of mass.

Mass-Balance Equations

In general, the calculations are rather complicated or complex. For instance, in the case of a dibasic acid, H^{2A} dissolved in water the two reactants can be specified as the conjugate base, A^{2-}, and the proton, H^+. The following equations of mass-balance could apply equally well to a base such as 1,2-diaminoethane, in which case the base itself is designated as the reactant A:

$$T_A = [A] + [HA] + [H_2A]$$
$$T_H = [H] + [HA] + 2[H_2A] - [OH]$$

With T_A the total concentration of species A. Note that it is customary to omit the ionic charges when writing and using these equations.

When the equilibrium constants are known and the total concentrations are specified there are two equations in two unknown "free concentrations"

[A] and [H]. This follows from the fact that $[HA] = \beta_1[A][H]$, $[H_2A] = \beta_2[A][H]^2$ and $[OH] = K_w[H]^{-1}$

$$T_A = [A] + \beta_1[A][H] + \beta_2[A][H]^2$$
$$T_H = [H] + \beta_1[A][H] + 2\beta_2[A][H]^2 - K_w[H]^{-1}$$

so the concentrations of the "complexes" are calculated from the free concentrations and the equilibrium constants. General expressions applicable to all systems with two reagents, A and B would be

$$T_A = [A] + \sum_i p_i\beta_i[A]^{p_i}[B]^{q_i}$$

$$T_B = [B] + \sum_i q_i\beta_i[A]^{p_i}[B]^{q_i}$$

It is easy to see how this can be extended to three or more reagents.

Polybasic Acids

The composition of solutions containing reactants A and H is easy to calculate as a function of p[H]. When [H] is known, the free concentration [A] is calculated from the mass-balance equation in A.

The diagram alongside, shows an example of the hydrolysis of the aluminium Lewis acid $Al^{3+}(aq)$ shows the species concentrations for a 5×10^{-6} M solution of an aluminium salt as a function of pH. Each concentration is shown as a percentage of the total aluminium.

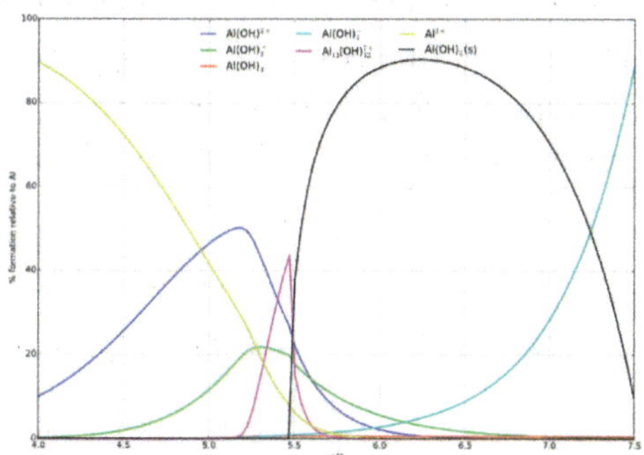

Species concentrations during hydrolysis of the aluminium.

Solution and Precipitation

The diagram above illustrates the point that a precipitate that is not one of the main species in the solution equilibrium may be formed. At pH just below 5.5 the main species present in a 5 μM solution of Al^{3+} are aluminium hydroxides $Al(OH)^{2+}$, $AlOH_2^+$ and $Al_{13}(OH)_{32}^{7+}$, but on raising the pH $Al(OH)_3$ precipitates from the solution. This occurs because $Al(OH)_3$ has a very large lattice energy. As the pH rises more and more $Al(OH)_3$ comes out of solution. This is an example of Le Châtelier's principle in action: Increasing the concentration of the hydroxide ion causes more aluminium hydroxide to precipitate, which removes hydroxide from the solution. When the hydroxide concentration becomes sufficiently high the soluble aluminate, $Al(OH)_4^-$, is former another common instance where precipitation occurs is when a metal cation interacts with an anionic ligand to form an electrically neutral complex. If the complex is hydrophobic, it will precipitate out of water. This occurs with the nickel ion Ni^{2+} and dimethylglyoxime, (dmgH2): in this case the lattice energy of the solid is not particularly large, but it greatly exceeds the energy of solvation of the molecule $Ni(dmgH)_2$.

Minimization of Free Energy

At equilibrium, G is at a minimum:

$$dG = \sum_{j=1}^{m} \mu_j \, dN_j = 0$$

For a closed system, no particles may enter or leave, although they may combine in various ways. The total number of atoms of each element will remain constant. This means that the minimization above must be subjected to the constraints:

$$\sum_{j=1}^{m} a_{ij} N_j = b_i^0$$

where a_{ij} is the number of atoms of element i in molecule j and b_j^0 is the total number of atoms of element i, which is a constant, since the system is closed. If there are a total of k types of atoms in the system, then there will be k such equations.

This is a standard problem in optimisation, known as constrained minimisation. The most common method of solving it is using the method of Lagrange multipliers, also known as undetermined multipliers (though other methods may be used).

Define:

$$\mathcal{G} = G + \sum_{i=1}^{k} \lambda_i \left(\sum_{j=1}^{m} a_{ij} N_j - b_i^0 \right) = 0$$

where the λ_i are the Lagrange multipliers, one for each element. This allows each of the N_j to be treated independently, and it can be shown using the tools of multivariate calculus that the equilibrium condition is given by

$$\frac{\partial \mathcal{G}}{\partial N_j} = 0 \quad \text{and} \quad \frac{\partial \mathcal{G}}{\partial \lambda_i} = 0$$

(For proof see Lagrange multipliers.)

This is a set of (m + k) equations in (m + k) unknowns (the N_j and the λ_i) and may, therefore, be solved for the equilibrium concentrations Nj as long as the chemical potentials are known as functions of the concentrations at the given temperature and pressure. (See Thermodynamic databases for pure substances.)

This method of calculating equilibrium chemical concentrations is useful for systems with a large number of different molecules. The use of k atomic element conservation equations for the mass constraint is straightforward, and replaces the use of the stoichiometric coefficient equations.

DETERMINATION OF EQUILIBRIUM CONSTANTS

Equilibrium constants are determined in order to quantify chemical equilibria. When an equilibrium constant K is expressed as a concentration quotient,

$$K = \frac{[\mathrm{S}]^{\sigma}[\mathrm{T}]^{\tau}\cdots}{[\mathrm{A}]^{\alpha}[\mathrm{B}]^{\beta}\cdots}$$

it is implied that the activity quotient is constant. For this assumption to be valid, equilibrium constants must be determined in a medium of relatively high ionic strength. Where this is not possible, consideration should be given to possible activity variation.

The equilibrium expression above is a function of the concentrations [A], [B] etc. of the chemical species in equilibrium. The equilibrium constant value can be determined if any one of these concentrations can be measured. The general procedure is that the concentration in question is measured for a series of solutions with known analytical concentrations of the reactants. Typically, a titration is performed with one or more reactants in the titration vessel and one or more reactants in the burette. Knowing the analytical concentrations of reactants initially in the reaction vessel and in the burette, all analytical concentrations can be derived as a function of the volume (or mass) of titrant added.

The equilibrium constants may be derived by best-fitting of the experimental data with a chemical model of the equilibrium system.

Experimental Methods

There are four main experimental methods. For less commonly used methods, see Rossotti and Rossotti.

Potentiometric Measurements

A free concentration [A] or activity {A} of a species A is measured by means of an ion selective electrode such as the glass electrode. If the electrode is calibrated using activity standards it is assumed that the Nernst equation applies in the form

$$E = E^0 + \frac{RT}{nF} \ln \{A\}$$

where E^0 is the standard electrode potential. When buffer solutions of known pH are used for calibration the meter reading will be a pH.

$$pH = \frac{nF}{RT}(E^0 - E)$$

At 298 K, 1 pH unit is approximately equal to 59 mV.

When the electrode is calibrated with solutions of known concentration, by means of a strong acid–strong base titration, for example, a modified Nernst equation is assumed.

$$E = E^0 + s \log_{10} [A]$$

where s is an empirical slope factor. A solution of known hydrogen ion concentration may be prepared by standardization of a strong acid against borax. Constant-boiling hydrochloric acid may also be used as a primary standard for hydrogen ion concentration.

Spectrophotometric Measurements

Absorbance

It is assumed that the Beer–Lambert law applies.

$$A = l \sum \varepsilon c$$

where l is the optical path length, ε is a molar absorbance at unit path length and c is a concentration. More than one of the species may contribute to the absorbance. In principle absorbance may be measured at one wavelength only, but in present-day practice it is common to record complete spectra.

Fluorescence (Luminescence) Intensity

It is assumed that the scattered light intensity is a linear function of species' concentrations.

$$I = \sum \varphi c$$

where φ is a proportionality constant.

NMR Chemical Shift Measurements

Chemical exchange is assumed to be rapid on the NMR time-scale. An individual chemical shift δ is the mole-fraction-weighted average of the shifts δ of nuclei in contributing species.

$$\bar{\delta} = \frac{\sum c_i \delta_i}{\sum c_i}$$

Example: the pK_a of the hydroxyl group in citric acid has been determined from ^{13}C chemical shift data to be 14.4. Neither potentiometry nor UV–visible spectrophotometry could be used for this determination.

Calorimetric Measurements

Simultaneous measurement of K and ΔH for 1:1 adducts is routinely carried out using isothermal titration calorimetry. Extension to more complex systems is limited by the availability of suitable software.

Range and Limitations

Potentiometry

The most widely used electrode is the glass electrode, which is selective for the hydrogen ion. This is suitable for all acid–base equilibria. log10 β values between about 2 and 11 can be measured directly by potentiometric titration using a glass electrode. This enormous range is possible because of the logarithmic response of the electrode. The limitations arise because the Nernst equation breaks down at very low or very high pH. The range can be extended by using the competition method. An example of the application of this method can be found in palladium(II) cyanide.

Absorbance and Luminescence

An upper limit on log10 β of 4 is usually quoted, corresponding to the precision of the measurements, but it also depends on how intense the effect is. Spectra of contributing species should be clearly distinct from each other

NMR

Limited precision of chemical shift measurements also puts an upper limit of about 4 on log10 β. Limited to diamagnetic systems. 1H NMR cannot be used with solutions of compounds in $^{1}H_2O$.

Calorimetry

Insufficient evidence is currently available.

Computational Methods

It is assumed that the collected experimental data comprise a set of data points. At each ith data point, the analytical concentrations of the reactants, $T_A(i)$, $T_B(i)$ etc. are known along with a measured quantity, y_i, that depends on one or more of these analytical concentrations. A general computational procedure has four main components:

- Definition of a chemical model of the equilibria
- Calculation of the concentrations of all the chemical species in each solution
- Refinement of the equilibrium constants
- Model selection

The Chemical Model

The chemical model consists of a set of chemical species present in solution, both the reactants added to the reaction mixture and the complex species formed from them. Denoting the reactants by A, B..., each complex species is specified by the stoichiometric coefficients that relate the particular combination of reactants forming them.

$$pA + qB \cdots \rightleftharpoons A_pB_q \cdots : \beta_{pq\cdots} = \frac{[A_pB_q \cdots]}{[A]^p[B]^q \cdots}$$

When using general-purpose computer programs, it is usual to use cumulative association constants, as shown above. Electrical charges are not shown in general expressions such as this and are often omitted from specific expressions, for simplicity of notation. In fact, electrical charges have no bearing on the equilibrium processes other that there being a requirement for overall electrical neutrality in all systems.

With aqueous solutions the concentrations of proton (hydronium ion) and hydroxide ion are constrained by the self-dissociation of water.

$$H_2O \rightleftharpoons H^+ + OH^- : K'_W = \frac{[H^+][OH^-]}{[H_2O]}$$

With dilute solutions the concentration of water is assumed constant, so the equilibrium expression is written in the form of the ionic product of water.

$$K_W = [H^+][OH^-]$$

When both H^+ and OH^- must be considered as reactants, one of them is eliminated from the model by specifying that its concentration be derived from the concentration of the other. Usually the concentration of the hydroxide ion is given by

$$[OH^-] = \frac{K_W}{[H^+]}$$

In this case the equilibrium constant for the formation of hydroxide has the stoichiometric coefficients −1 in regard to the proton and zero for the other reactants. This has important implications for all protonation equilibria in aqueous solution and for hydrolysis constants in particular.

It is quite usual to omit from the model those species whose concentrations are considered negligible. For example, it is usually assumed then there is no interaction between the reactants and/or complexes and the electrolyte used to maintain constant ionic strength or the buffer used to maintain constant pH. These assumptions may or may not be justified. Also, it is implicitly assumed that there are no other complex species present. When complexes are wrongly ignored a systematic error is introduced into the calculations.

Equilibrium constant values are usually estimated initially by reference to data sources.

Speciation Calculations

A speciation calculation is one in which concentrations of all the species in an equilibrium system are calculated, knowing the analytical concentrations, T_A, T_B etc. of the reactants A, B etc. This means solving a set of nonlinear equations of mass-balance

$$T_{\mathrm{A}} = [\mathrm{A}] + \sum p\beta_{pq\cdots}[\mathrm{A}]^p[\mathrm{B}]^q\cdots$$

$$T_{\mathrm{B}} = [\mathrm{B}] + \sum q\beta_{pq\cdots}[\mathrm{A}]^p[\mathrm{B}]^q\cdots$$

$$\vdots$$

for the free concentrations [A], [B] etc. The concentrations of the complexes are derived from the free concentrations via the chemical model. Some authors include the free reactant terms in the sums by declaring identity (unit) β constants for which the stoichiometric coefficients are 1 for the reactant concerned and zero for all other reactants:

$$[\mathrm{A}]=\beta_{10\ldots}[\mathrm{A}], [\mathrm{B}]=\beta_{01\ldots}[\mathrm{B}]\ldots$$
$$\beta_{10\ldots}=\beta_{01\ldots}\cdots=1$$

In this manner, all chemical species, including the free reactants, are treated in the same way, having been formed from the combination of reactants that is specified by the stoichiometric coefficients. The mass-balance equations assume the simpler form.

$$T_{\mathrm{A}} = \sum p\beta_{pq\cdots}[\mathrm{A}]^p[\mathrm{B}]^q\cdots$$

$$T_{\mathrm{B}} = \sum q\beta_{pq\cdots}[\mathrm{A}]^p[\mathrm{B}]^q\cdots$$

$$\vdots$$

In a titration system the analytical concentrations of the reactants at each titration point are obtained from the initial conditions, the burette concentrations and volumes. The analytical (total) concentration of a reactant R at the ith titration point is given by

$$T_{\mathrm{R}} = \frac{\mathrm{R}_0 + v_i[\mathrm{R}]}{v_0 + v_i}$$

where R_0 is the initial amount of R in the titration vessel, v0 is the initial volume, [R] is the concentration of R in the burette and vi is the volume added. The burette concentration of a reactant not present in the burette is taken to be zero.

In general, solving these nonlinear equations presents a formidable challenge because of the huge range over which the free concentrations may vary. At the beginning, values for the free concentrations must be estimated. Then, these values are refined, usually by means of Newton–Raphson iterations. The logarithms of the free concentrations may be refined rather than the free concentrations themselves. Refinement of the logarithms of the free concentrations has the added advantage of automatically imposing a non-negativity constraint on the free concentrations. Once the free reactant concentrations have been calculated, the concentrations of the complexes are derived from them and the equilibrium constants.

Note that the free reactant concentrations can be regarded as implicit parameters in the equilibrium constant refinement process. In that context the values of the free concentrations are constrained by forcing the conditions of mass-balance to apply at all stages of the process.

Equilibrium Constant Refinement

The objective of the refinement process is to find equilibrium constant values that give the best fit to the experimental data. This is usually achieved by minimising an objective function, U, by the method of non-linear least-squares. First the residuals are defined as

$$r_i = y_i^{\mathrm{obs}} - y_i^{\mathrm{calc}}$$

Then the most general objective function is given by

$$U = \sum_i \sum_j r_i W_{ij} r_j$$

The matrix of weights, W, should be, ideally, the inverse of the variance-covariance matrix of the observations. It is rare for this to be known. However, when it is, the expectation value of U is one, which means that the data are fitted within experimental error. Most often only the diagonal elements are known, in which case the objective function simplifies to

$$U = \sum_i W_{ii} r_i^2$$

with $W_{ij} = 0$ when $j \neq i$. Unit weights, $W_{ii} = 1$, are often used but, in that case, the expectation value of U is the root mean square of the experimental errors.

The minimization may be performed using the Gauss–Newton method. Firstly the objective function is linearised by approximating it as a first-order Taylor series expansion about an initial parameter set, p.

$$U = U^0 + \sum_i \frac{\partial U}{\partial p_i} \delta p_i$$

The increments δ_{pi} are added to the corresponding initial parameters such that U is less than U^0. At the minimum the derivatives $\frac{\partial U}{\partial p_i}$, which are simply related to the elements of the Jacobian matrix, J

$$J_{jk} = \frac{\partial y_j^{calc}}{\partial p_k}$$

where p_k is the kth parameter of the refinement, are equal to zero. One or more equilibrium constants may be parameters of the refinement. However, the measured quantities (see above) represented by y are not expressed in terms of the equilibrium constants, but in terms of the species concentrations, which are implicit functions of these parameters. Therefore, the Jacobian elements must be obtained using implicit differentiation.

The parameter increments δp are calculated by solving the normal equations, derived from the conditions that $\frac{\partial \mathrm{U}}{\partial \mathrm{p}} = 0$ at the minimum.

$$(\mathbf{J}^T\mathbf{W}\mathbf{J})\,\delta\mathbf{p} = \mathbf{J}^T\mathbf{W}\mathbf{r}$$

The increments $\delta\mathbf{p}$ are added iteratively to the parameters

$$\mathbf{p}^{n+1} = \mathbf{p}^n + \delta\mathbf{p}$$

where n is an iteration number. The species concentrations and y^{calc} values are recalculated at every data point. The iterations are continued until no significant reduction in U is achieved, that is, until a convergence criterion is satisfied. If, however, the updated parameters do not result in a decrease of the objective function, that is, if divergence occurs, the increment calculation must be modified. The simplest modification is to use a fraction, f, of calculated increment, so-called shift-cutting.

$$\mathbf{p}^{n+1} = \mathbf{p}^n + f\delta\mathbf{p}$$

In this case, the direction of the shift vector, δp, is unchanged. With the more powerful Levenberg–Marquardt algorithm, on the other hand, the shift vector is rotated towards the direction of steepest descent, by modifying the normal equations,

$$(\mathbf{J}^T\mathbf{W}\mathbf{J} + \lambda\mathbf{I})\,\delta\mathbf{p} = \mathbf{J}^T\mathbf{W}\mathbf{r}$$

where λ is the Marquardt parameter and I is an identity matrix. Other methods of handling divergence have been proposed.

A particular issue arises with NMR and spectrophotometric data. For the latter, the observed quantity is absorbance, A, and the Beer–Lambert law can be written as

$$A_i = \sum \varepsilon_{pq\cdots} c_{pq\cdots,i}$$

It can be seen that absorbance, A, is a linear function of the molar absorbptivities, ε, at the path length used. In matrix notation

$$\mathbf{A} = \boldsymbol{\varepsilon}\mathbf{C}$$

There are two approaches to the calculation of the unknown molar absorptivities

The ε values are considered parameters of the minimization and the Jacobian is constructed on that basis. However, the ε values themselves are calculated at each step of the refinement by linear least-squares:

$$\boldsymbol{\varepsilon} = (\mathbf{C}^T\mathbf{C})^{-1}\mathbf{C}^T\mathbf{A}$$

using the refined values of the equilibrium constants to obtain the speciation. The matrix

$$(\mathbf{C}^T\mathbf{C})^{-1}\mathbf{C}^T$$

is an example of a pseudo-inverse.

- The Beer–Lambert law is written as

$$\mathbf{A} = \left((\mathbf{C}^T\mathbf{C})^{-1}\mathbf{C}^T\mathbf{A}\right)\mathbf{C}$$

Golub and Pereyra showed how the pseudo-inverse can be differentiated so that parameter increments for both molar absorptivities and equilibrium constants can be calculated by solving the normal equations.

Parameter Errors and Correlation

In the region close to the minimum of the objective function, U, the system approximates to a linear least-squares system, for which

$$\mathbf{p} = (\mathbf{J}^T\mathbf{W}\mathbf{J})^{-1}\mathbf{J}^T\mathbf{W}\mathbf{y}^{obs}$$

Therefore, the parameter values are (approximately) linear combinations of the observed data values and the errors on the parameters, p, can be obtained by error propagation from the observations, y^{obs}, using the linear formula. Let the variance-covariance matrix for the observations be denoted by $\boldsymbol{\Sigma}^y$ and that of the parameters by $\boldsymbol{\Sigma}^p$. Then,

$$\boldsymbol{\Sigma}^p = (\mathbf{J}^T\mathbf{W}\mathbf{J})^{-1}\mathbf{J}^T\mathbf{W}\boldsymbol{\Sigma}^y\mathbf{W}^T\mathbf{J}(\mathbf{J}^T\mathbf{W}\mathbf{J})^{-1}$$

When $\mathbf{W} = (\boldsymbol{\Sigma}^y)^{-1}$, this simplifies to

$$\boldsymbol{\Sigma}^p = (\mathbf{J}^T\mathbf{W}\mathbf{J})^{-1}$$

In most cases the errors on the observations are un-correlated, so that Σ^{y} is diagonal. If so, each weight should be the reciprocal of the variance of the corresponding observation. For example, in a potentiometric titration, the weight at a titration point, k, can be given by

$$W_k = \frac{1}{\sigma_E^2 + \left(\frac{\partial E}{\partial v}\right)_k^2 \sigma_v^2}$$

where σ_E is the error in electrode potential or pH $\left(\frac{\partial E}{\partial v}\right)_k$ is the slope of the titration curve and σ_v is the error on added volume.

When unit weights are used $(\mathbf{W} = \mathbf{I}, \mathbf{p} = (\mathbf{J}^T\mathbf{J})^{-1}\mathbf{J}^T\mathbf{y})$ it is implied that the experimental errors are uncorrelated and all equal: $\Sigma^y = \sigma^2 I$, where σ^2 is known as the variance of an observation of unit weight, and I is an identity matrix. In this case σ^2 is approximated by

$$\sigma^2 = \frac{U}{n_d - n_p}$$

where U is the minimum value of the objective function and n_d and n_p are the number of data and parameters, respectively.

$$\mathbf{\Sigma^p} = \frac{U}{n_d - n_p}(\mathbf{J}^T\mathbf{J})^{-1}$$

In all cases, the variance of the parameter pi is given by Σ^p_{ii} and the covariance between parameters pi and pj is given by Σ^p_{ij}. Standard deviation is the square root of variance. These error estimates reflect only random errors in the measurements. The true uncertainty in the parameters is larger due to the presence of systematic errors—which, by definition, cannot be quantified.

Note that even though the observations may be uncorrelated, the parameters are always correlated.

Derived Constants

When cumulative constants have been refined it is often useful to derive stepwise constants from them. The general procedure is to write down the defining expressions for all the constants involved and then to equate concentrations. For example, suppose that one wishes to derive the pKa for removing one proton from a tribasic acid, LH_3, such as citric acid.

$$L^{3-} + 2H^+ \rightleftharpoons LH_2^-; \quad [LH_2^-]=\beta_{12}[L^{3-}][H^+]^2$$

$$L^{3-} + 3H^+ \rightleftharpoons LH_3; \quad [LH_3]=\beta_{13}[L^{3-}][H^+]^3$$

The stepwise *association* constant for formation of LH_3 is given by

$$LH_2^- + H^+ \rightleftharpoons LH_3; \quad [LH_3] = K[LH_2^-][H^+]$$

Substitute the expressions for the concentrations of LH_3 and LH_2^- into this equation

$$\beta_{13}[L^{3-}][H^+]^3 = K\beta_{12}[L^{3-}][H^+]^2[H^+]$$

whence

$$\beta_{13} = K\beta_{12}; K = \frac{\beta_{13}}{\beta_{12}}$$

and since $pK_a = -\log_{10}\frac{1}{K}$ its value is given by

$$pK_{a1} = \log_{10}\beta_{13} - \log_{10}\beta_{12}$$

When calculating the error on the stepwise constant, the fact that the cumulative constants are correlated must accounted for. By error propagation

$$\sigma_K^2 = \sigma_{\beta_{12}}^2 + \sigma_{\beta_{13}}^2 - 2\sigma_{\beta_{12}}\sigma_{\beta_{13}}\rho_{12,13}$$

and

$$\sigma_{\log_{10} K} = \frac{\sigma_K}{K}$$

Model Selection

Once a refinement has been completed the results should be checked to verify that the chosen model is acceptable. generally speaking, a model is acceptable when the data are fitted within experimental error, but there is no single criterion to use to make the judgement. The following should be considered.

The Objective Function

When the weights have been correctly derived from estimates of experimental error, the expectation value of $\frac{U}{n_d - n_p}$ is 1. It is therefore very useful to estimate experimental errors and derive some reasonable weights from them as this is an absolute indicator of the goodness of fit.

When unit weights are used, it is implied that all observations have the same variance. $\frac{U}{n_d - n_p}$ is expected to be equal to that variance.

Parameter Errors

One would want the errors on the stability constants to be roughly commensurate with experimental error. For example, with pH titration data, if pH is measured to 2 decimal places, the errors of log10 β should not be much larger than 0.01. In exploratory work where the nature of the species present is not known in advance, several different chemical models may be tested and compared. There will be models where the uncertainties in the best estimate of an equilibrium constant may be somewhat or even significantly larger than σpH, especially with those constants governing the formation of comparatively minor species, but the decision as to how large is acceptable remains subjective. The decision process as to whether or not to include comparatively uncertain equilibria in a model, and for the comparison of competing models in general, can be made objective and has been outlined by Hamilton.

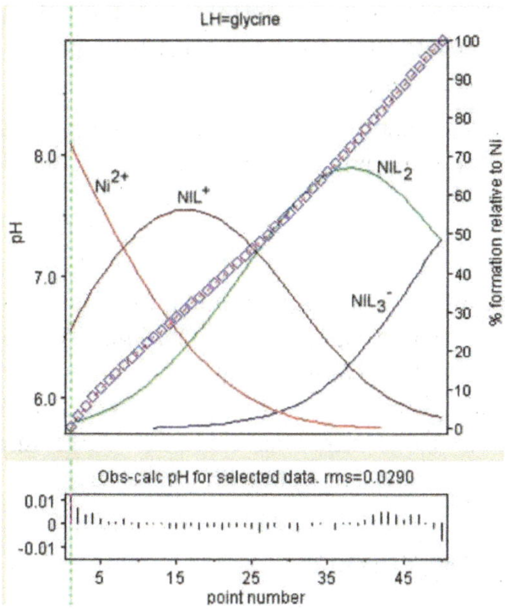

Distribution of Residuals

At the minimum in U the system can be approximated to a linear one, the residuals in the case of unit weights are related to the observations by

$$\mathbf{r} = \mathbf{y}^{obs} - \mathbf{J}(\mathbf{J}^T\mathbf{T})^{-1}\mathbf{J}^T\mathbf{y}^{obs}$$

The symmetric, idempotent matrix $J(J^TT)^{-1}J$ is known in the statistics literature as the hat matrix, H.

Thus,

$$\mathbf{r} = (\mathbf{I} - \mathbf{H})\,\mathbf{y}^{obs}$$

and

$$\mathbf{M}^{r} = (\mathbf{I} - \mathbf{H})\,\mathbf{M}^{y}\,(\mathbf{I} - \mathbf{H})$$

where I is an identity matrix and M_r and M_y are the variance-covariance matrices of the residuals and observations, respectively. This shows that even though the observations may be uncorrelated, the residuals are always correlated.

The diagram at the right shows the result of a refinement of the stability constants of $Ni(Gly)^+$, $Ni(Gly)_2$ and $Ni(Gly)_3^-$ (where GlyH = glycine). The observed values are shown a blue diamonds and the species concentrations, as a percentage of the total nickel, are superimposed. The residuals are shown in the lower box. The presence of correlation is evident in the way sequences all have the same sign. Correlation notwithstanding, the magnitudes of the residuals show some randomness. Individual residuals are mostly commensurate with experimental error (about 0.002 in pH). This is about as good as it gets.

Physical Constraints

Some physical constraints are usually incorporated in the calculations. For example, all the concentrations of free reactants and species must have positive values and association constants must have positive values.

With spectrophotometric data the molar absorptivity (or emissivity) values should all be positive. Most computer programs do not impose this constraint on the calculations.

Other Models

If the model is not acceptable, a variety of other models should be examined to find one that best fits the experimental data, within experimental error. The main difficulty is with the so-called minor species. These are species whose concentration is so low that the effect on the measured quantity is at or below the level of error in the experimental measurement. The constant for a minor species may prove impossible to determine if there is no means to increase the concentration of the species. .

EQUILIBRIUM CONSTANT

The equilibrium constant of a chemical reaction is the value of the reaction quotient when the reaction has reached equilibrium. An equilibrium constant value is independent of the analytical concentrations of the reactant and product species in a mixture, but depends on temperature and on ionic strength. Known equilibrium constant values can be used to determine the composition of a system at equilibrium.

For a general chemical equilibrium

$$\alpha A + \beta B \ldots \rightleftharpoons \rho R + \sigma S \ldots$$

the thermodynamic equilibrium constant can be defined such that, at equilibrium,

$$K^{\ominus} = \frac{\{R\}^{\rho}\{S\}^{\sigma}\ldots}{\{A\}^{\alpha}\{B\}^{\beta}\ldots}$$

where curly brackets denote the thermodynamic activities of the chemical species. The right-hand side of this equation corresponds to the reaction quotient Q for arbitrary values of the activities, and becomes the equilibrium constant as shown when the reaction is at equilibrium.

An equilibrium constant is related to the standard Gibbs free energy change for the reaction.

$$\Delta G^{\ominus} = -RT \ln K^{\ominus}$$

If deviations from ideal behaviour are neglected, the activities of solutes may be replaced by concentrations, [A], and the activity quotient becomes a concentration quotient, K_c.

$$K_c = \frac{[R]^{\rho}[S]^{\sigma}\ldots}{[A]^{\alpha}[B]^{\beta}\ldots}$$

K_c is defined as equal to the thermodynamic equilibrium constant but with concentrations of reactants and products instead of activities. (K_c appears here to have units of concentration raised to some power while K° is dimensionless; however as discussed below under Definitions, the concentration factors in K_c are properly divided by a standard concentration so that K_c is dimensionless also.)

Again assuming ideal behavior, the activity of a solvent may be replaced by its mole fraction, or approximately by 1 in dilute solution. The activity of a pure liquid or solid phase is exactly 1. The activity of a species in an ideal gas phase may be replaced by its partial pressure.

A knowledge of equilibrium constants is essential for the understanding of many chemical systems, as well as biochemical processes such as oxygen transport by hemoglobin in blood and acid-base homeostasis in the human body.

Stability constants, formation constants, binding constants, association constants and dissociation constants are all types of equilibrium constant. See also Determination of equilibrium constants for experimental and computational methods.

Types of Equilibrium Constants

Cumulative and Stepwise Formation Constants

A cumulative or overall constant, given the symbol β, is the constant for the formation of a complex from reagents. For example, the cumulative constant for the formation of ML_2 is given by

$$M + 2L \rightleftharpoons ML_2; \quad [ML_2] = \beta_{12}[M][L]^2$$

The stepwise constant, K, for the formation of the same complex from ML and L is given by

$$ML + L \rightleftharpoons ML_2; \quad [ML_2] = K[ML][L] = K\beta_{11}[M][L]^2$$

It follows that

$$\beta_{12} = K\beta_{11}$$

A cumulative constant can always be expressed as the product of stepwise constants.

There is no agreed notation for stepwise constants, though a symbol such as K_{ML}^{L} is sometimes found in the literature. It is best always to define each stability constant by reference to an equilibrium expression.

Competition Method

A particular use of a stepwise constant is in the determination of stability constant values outside the normal range for a given method. For example, EDTA complexes of many metals are outside the range for the potentiometric method. The stability constants for those complexes were determined by competition with a weaker ligand.

$$\mathrm{ML + L' \rightleftharpoons ML' + L} \quad [\mathrm{ML'}] = K\frac{[\mathrm{ML}][\mathrm{L'}]}{[\mathrm{L}]} = K\frac{\beta_{\mathrm{ML}}[\mathrm{M}][\mathrm{L}][\mathrm{L'}]}{[\mathrm{L}]} = K\beta_{\mathrm{ML}}[\mathrm{M}][\mathrm{L'}]; \quad \beta_{\mathrm{ML'}} = K\beta_{\mathrm{ML}}$$

The formation constant of $[Pd(CN)_4]^{2-}$ was determined by the competition method.

Association and Dissociation Constants

In organic chemistry and biochemistry it is customary to use pK_a values for acid dissociation equilibria.

$$\mathrm{p}K_{\mathrm{a}} = -\log K_{\mathrm{diss}} = \log\left(\frac{1}{K_{\mathrm{diss}}}\right)$$

Where K_{diss} is a stepwise acid dissociation constant. For bases, the base association constant, pK_b is used. For any given acid or base the two constants are related by $pK_a + pK_b = pK_w$, so pK_a can always be used in calculations.

On the other hand, stability constants for metal complexes, and binding constants for host–guest complexes are generally expressed as association constants. When considering equilibria such as

$$\mathrm{M + HL \rightleftharpoons ML + H}$$

it is customary to use association constants for both ML and HL. Also, in generalized computer programs dealing with equilibrium constants it is general practice to use cumulative constants rather than stepwise constants and to omit ionic charges from equilibrium expressions. For example, if NTA, nitrilotriacetic acid, $N(CH_2CO_2H)_3$ is designated as H_3L and forms complexes ML and MHL with a metal ion M, the following expressions would apply for the dissociation constants.

$$H_3L \rightleftharpoons H_2L + H; \quad pK_1 = -\log\left(\frac{[H_2L][H]}{[H_3L]}\right)$$

$$H_2L \rightleftharpoons HL + H; \quad pK_2 = -\log\left(\frac{[HL][H]}{[H_2L]}\right)$$

$$HL \rightleftharpoons L + H; \quad pK_3 = -\log\left(\frac{[L][H]}{[HL]}\right)$$

The cumulative association constants can be expressed as

$$L + H \rightleftharpoons HL; \quad \log\beta_{011} = \log\left(\frac{[HL]}{[L][H]}\right) = pK_3$$

$$L + 2H \rightleftharpoons H_2L; \quad \log\beta_{012} = \log\left(\frac{[H_2L]}{[L][H]^2}\right) = pK_3 + pK_2$$

$$L + 3H \rightleftharpoons H_3L; \quad \log\beta_{013} = \log\left(\frac{[H_3L]}{[L][H]^3}\right) = pK_3 + pK_2 + pK_1$$

$$M + L \rightleftharpoons ML; \quad \log\beta_{110} = \log\left(\frac{[ML]}{[M][L]}\right)$$

$$M + L + H \rightleftharpoons MLH; \quad \log\beta_{111} = \log\left(\frac{[MLH]}{[M][L][H]}\right)$$

Note how the subscripts define the stoichiometry of the equilibrium product.

Micro-Constants

When two or more sites in an asymmetrical molecule may be involved in an equilibrium reaction there are more than one possible equilibrium constants. For example, the molecule L-DOPA has two non-equivalent hydroxyl groups which may be deprotonated. Denoting L-DOPA as LH_2, the following diagram shows all the species that may be formed (X = $CH_2CH(NH_2)CO_2H$):

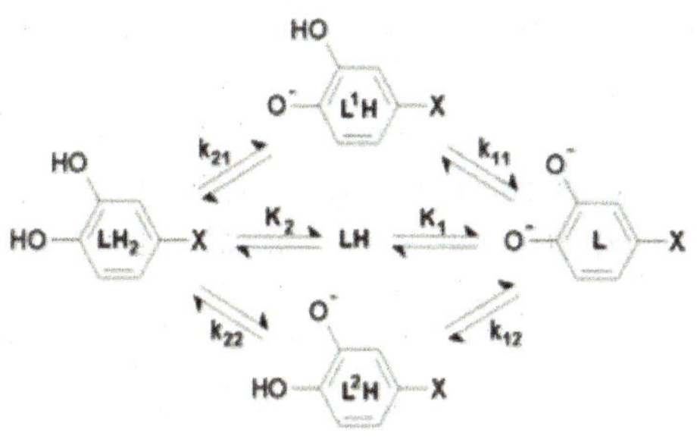

The first protonation constants are

$$[L^1H] = k_{11}[L][H], \; [L^2H] = k_{12}[L][H]$$

The concentration of LH^- is the sum of the concentrations of the two micro-species. Therefore, the equilibrium constant for the reaction, the macro-constant, is the sum of the micro-constants.

$$K_1 = k_{11} + k_{12}$$

In the same way,

$$K_2 = k_{21} + k_{22}$$

Lastly, the cumulative constant is

$$\beta_2 = K_1K_2 = k_{11}k_{21} = k_{12}k_{22}$$

Thus, although there are seven micro-and macro-constants, only three of them are mutually independent. Moreover, the isomerization constant, K_i, is equal to the ratio of the microconstants.

$$K_i = k_{11}/k_{12}$$

In L-DOPA the isomerization constant is 0.9, so the micro-species L^1H and L^2H have almost equal concentrations at all pH values.

In general a macro-constant is equal to the sum of all the micro-constants and the occupancy of each site is proportional to the micro-constant. The site of protonation can be very important, for example, for biological activity.

Micro-constants cannot be determined individually by the usual methods, which give macro-constants. Methods which have been used to determine micro-constants include:

- blocking one of the sites, for example by methylation of a hydroxyl group, to determine one of the micro-constants
- using a spectroscopic technique, such as infrared spectroscopy, where the different micro-species give different signals.
- applying mathematical procedures to 13C NMR data.

pH Considerations (Brønsted Constants)

pH is defined in terms of the activity of the hydrogen ion

$$pH = -\log_{10}\{H^+\}$$

In the approximation of ideal behaviour, activity is replaced by concentration. pH is measured by means of a glass electrode, a mixed equilibrium constant, also known as a Brønsted constant, may result.

$$HL \rightleftharpoons L + H; \quad pK = -\log\left(\frac{[L]\{H\}}{[HL]}\right)$$

It all depends on whether the electrode is calibrated by reference to solutions of known activity or known concentration. In the latter case the equilibrium constant would be a concentration quotient. If the electrode is calibrated in terms of known hydrogen ion concentrations it would be better to write p[H] rather than pH, but this suggestion is not generally adopted.

Hydrolysis Constants

In aqueous solution the concentration of the hydroxide ion is related to the concentration of the hydrogen ion by

$$K_W=[H][OH]$$
$$[OH]=K_W[H]^{-1}$$

The first step in metal ion hydrolysis can be expressed in two different ways

$$\begin{cases} M(H_2O) \rightleftharpoons M(OH)+H; & [M(OH)]=\beta^*[M][H]^{-1} \\ M+OH \rightleftharpoons M(OH); & [M(OH)]=K[M][OH]=KK_W[M][H]^{-1} \end{cases}$$

It follows that $\beta^* = KK_W$. Hydrolysis constants are usually reported in the β^* form and this leads to them appearing to have strange values. For example, if $\log K = 4$ and $\log K_W = -14$, $\log \beta^* = 4 + (-14) = -10$. In general when the hydrolysis product contains n hydroxide groups $\lg \beta^* = \log K + n \log K_W$

Conditional Constants

Conditional constants, also known as apparent constants, are concentration quotients which are not true equilibrium constants but can be derived from them. A very common instance is where pH is fixed at a particular value. For example, in the case of iron(III) interacting with EDTA, a conditional constant could be defined by

$$K_{cond} = \frac{[\text{Total Fe bound to EDTA}]}{[\text{Total Fe not bound to EDTA}] \times [\text{Total EDTA not bound to Fe}]}$$

This conditional constant will vary with pH. It has a maximum at a certain pH. That is the pH where the ligand sequesters the metal most effectively.

In biochemistry equilibrium constants are often measured at a pH fixed by means of a buffer solution. Such constants are, by definition, conditional and different values may be obtained when using different buffers.

Gas-Phase Equilibria

For equilibria in a gas phase, fugacity, f, is used in place of activity. However, fugacity has the dimension of pressure, so it must be divided by a standard pressure, usually 1 bar, in order to produce a dimensionless quantity, $\frac{f}{p^\ominus}$. An equilibrium constant is expressed in terms of the dimensionless quantity. For example, for the equilibrium $2NO_2 \rightleftharpoons N_2O_4$,

$$\frac{f_{N_2O_4}}{p^\ominus} = K\left(\frac{f_{NO_2}}{p^\ominus}\right)^2$$

Fugacity is related to partial pressure, p, by a dimensionless fugacity coefficient ϕ: $f = \phi p$. Thus, for the example,

$$K = \frac{\phi_{N_2O_4} p_{N_2O_4}/p^\ominus}{(\phi_{NO_2} p_{NO_2}/p^\ominus)^2}$$

Usually the standard pressure is omitted from such expressions. Expressions for equilibrium constants in the gas phase then resemble the expression for solution equilibria with fugacity coefficient in place of activity coefficient and partial pressure in place of concentration.

$$K = \frac{\phi_{N_2O_4} p_{N_2O_4}}{(\phi_{NO_2} p_{NO_2})^2}$$

Thermodynamics

Definitions

For a general chemical equilibrium

$$\alpha A + \beta B \ldots \rightleftharpoons \rho R + \sigma S \ldots$$

the thermodynamic definition of the equilibrium constant is that, at equilibrium,

$$K^{\ominus} = \frac{\{R\}^{\rho}\{S\}^{\sigma}\ldots}{\{A\}^{\alpha}\{B\}^{\beta}\ldots}$$

where {A} is the activity, at equilibrium, of the chemical species A, etc. It is conventional to put the activities of the products in the numerator and those of the reactants in the denominator. Since activity is a dimensionless quantity, the thermodynamic equilibrium constant, K°, is also a dimensionless quantity. This is important because K° is related to the standard free energy change, ΔG°, for the reaction.

$$\Delta G^{\ominus} = -RT \ln K^{\ominus}$$

R is the gas constant and T the absolute or thermodynamic temperature. Taking the logarithm of K° is only possible when K°has no dimensions, that is, when it is a pure (positive) number. However it is often more convenient to use concentrations (or partial pressures in the gas phase) than activities, so the equilibrium expression may be written as the product of three components, a reaction quotient of equilibrium concentrations, Q^E, a quotient of activity factors, Γ, and a quotient of standard concentrations, C^0.

$$K^{\ominus} = \frac{[R]^{\rho}[S]^{\sigma}\ldots}{[A]^{\alpha}[B]^{\beta}\ldots} \times \frac{\gamma_R{}^{\rho}\gamma_S{}^{\sigma}\ldots}{\gamma_A{}^{\alpha}\gamma_B{}^{\beta}\ldots} \times \frac{(c_A^{\ominus})^{\alpha}(c_B^{\ominus})^{\beta}\ldots}{(c_R^{\ominus})^{\rho}(c_S^{\ominus})^{\sigma}\ldots} = Q^E \Gamma C^0$$

The symbol [A] stands for a generalized concentration or partial pressure of the chemical species A, The activity coefficients, γ, are dimensionless and $c_A^{\ominus}$ stands for the standard concentration or standard partial pressure. By convention the standard concentrations have a value of 1 mol/dm³ (or 1 atm), so the third factor, C^0, has a numerical value of 1 and units that are the reciprocal of the units of the concentration quotient. K° remains dimensionless. so the expression simplifies to

$$K^{\ominus} = Q^{\mathrm{E}} \times \Gamma \times 1 = Q^{\mathrm{E}}\Gamma$$

It follows that the "Concentration constant", K_c, is related to the thermodynamic equilibrium constant by

$$K_c = \frac{[\mathrm{R}]^{\rho}[\mathrm{S}]^{\sigma}\ldots}{[\mathrm{A}]^{\alpha}[\mathrm{B}]^{\beta}\ldots} \times C^0 = \frac{K^{\ominus}}{\Gamma}$$

Note that when K_c is defined in this way it has no dimension, so log K_c is permissible.

When an equilibrium constant value is to be determined, there are three options for dealing with the activity coefficients.

- Use known or calculated activity coefficients, together with concentrations of reactants. For equilibria in solution estimates of the activity coefficients of charged species can be obtained using Debye–Hückel theory, an extended version, or SIT theory. For uncharged species, the activity coefficient γ_0 mostly follows a "salting-out" model: $\log_{10} \gamma_0 = bI$ where I stands for ionic strength.
- Assume that the activity coefficients are all equal to 1. This is acceptable when all species are uncharged and the solvent has low polarity.
- For equilibria in solution use a medium of high ionic strength. In effect this redefines the standard state as referring to the medium. Activity coefficients in the standard state are, by definition, equal to 1. The value of an equilibrium constant determined in this manner is dependent on the ionic strength. When published constants refer to an ionic strength other than the one required for a particular application, they may be adjusted by means of specific ion theory (SIT) and other theories.

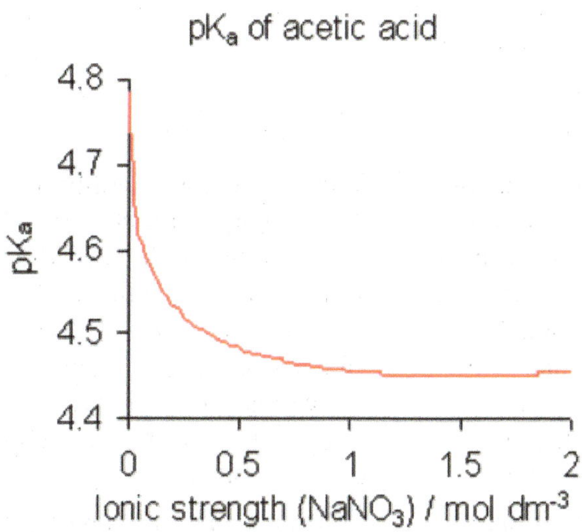

Variation of log K_c of acetic acid with ionic strength

Dimensionality

Any equilibrium constant, K, must be dimensionless. An equilibrium constant is related to the standard Gibbs free energy change for the reaction by the expression

$$\Delta G^{\ominus} = -RT \ln K$$

This requires that K be a pure number from which a logarithm can be derived. In the case of a simple equilibrium

$$A + B \rightleftharpoons AB$$

the thermodynamic equilibrium constant is defined in terms of the activities, {AB}, {A} and {B}, of the species in equilibrium with each other.

$$K = \frac{\{AB\}}{\{A\}\{B\}}$$

As activity is a dimensionless quantity, K is dimensionless.Now, each activity term can be expressed as a product of a concentration [X] and a orresponding activity coefficient, $\gamma(X)$. Therefore

$$K = \frac{[AB]}{[A][B]} \times \frac{\gamma(AB)}{\gamma(A)\gamma(B)}$$

Since each activity coefficient has the dimension 1/concentration, K remains dimensionless. However, when the quotient of activity coefficients is set equal to 1, K appears to have the dimension of 1/concentration.

$$K = \frac{[AB]}{[A][B]}$$

This is what usually happens in practice when an equilibrium constant is calculated as a quotient of concentration values. The assumption underlying this practice is that the quotient of activities is constant under the conditions in which the equilibrium constant value is determined. There conditions are usually achieved by keeping the reaction temperature constant and by using a medium of relatively high ionic strength as the solvent. It is not unusual, particularly in texts relating to biochemical equilibria, to see a value quoted with a dimension. The justification for this practice is that the concentration scale used may be either mol dm^{-3} or mmol dm^{-3}, so that the concentration unit has to be stated in order to avoid there being any ambiguity.

In general equilibria between two reagents can be expressed as

$$pA + qB \rightleftharpoons A_pB_q$$

in which case the equilibrium constant is defined, in terms of numerical concentration values, as

$$K = \frac{[A_pB_q]}{[A]^p[B]^q}$$

The apparent dimension of this K value is concentration^{1-p-q}; this may be written as $M^{(1-p-q)}$ or $mM^{(1-p-q)}$, where the symbol M signifies a molar concentration (1 M = 1 mol dm^{-3})

A further complications arises from the fact that an equilibrium constant can be defined as an association constant or as a dissociation constant. For dissociation constants

$$A_pB_q \rightleftharpoons pA + qB;\ K = \frac{[A]^p[B]^q}{[A_pB_q]}$$

The apparent dimension associated with a dissociation constant is the reciprocal of the dimension associated with the corresponding association constant.

Derivation of Equilibrium Constant Expression

Temperature, T, is assumed to be constant in the following treatment.

Thermodynamic equilibrium is characterized by the free energy for the whole (closed) system being a minimum. For systems at constant pressure the Gibbs free energy is minimum. The slope of the reaction free energy, δG_r with respect to the reaction coordinate, ξ, is zero when the free energy is at its minimum value.

$$\delta G_r = \left(\frac{\partial G}{\partial \xi}\right)_{T,P};\quad \delta G_r(\mathrm{Eq}) = 0$$

The free energy change, δG_r, can be expressed as a weighted sum of chemical potentials, the partial molar free energies of the species in equilibrium. The potential, μ_i, of the ith species in a chemical reaction is the partial derivative of the free energy with respect to the number of moles of that species, N_i

$$\mu_i = \left(\frac{\partial G}{\partial N_i}\right)_{T,P}$$

A general chemical equilibrium can be written as

$$\sum_j n_j \mathrm{Reactant}_j \rightleftharpoons \sum_k m_k \mathrm{Product}_k \qquad \approx \alpha A + \beta B \ldots \rightleftharpoons \rho R + \sigma S \ldots$$

where n_j are the stoichiometric coefficients of the reactants in the equilibrium equation, and m_j are the coefficients of the products. The value of δG_r for these reactions is a function of the chemical potentials of all the species.

$$\delta G_r = \sum_k m_k\mu_k - \sum_j n_j\mu_j$$

The chemical potential, μ_i, of the ith species can be calculated in terms of its activity, a_i.

$$\mu_i = \mu_i^{\ominus} + RT\ln a_i$$

$\mu_i^{\ominus}$ is the standard chemical potential of the species, R is the gas constant and T is the temperature. Setting the sum for the reactants j to be equal to the sum for the products, k, so that $\delta G_r(Eq) = 0$

$$\sum_j n_j(\mu_j^{\ominus} + RT\ln a_j) = \sum_k m_k(\mu_k^{\ominus} + RT\ln a_k)$$

Rearranging the terms,

$$\sum_k m_k\mu_k^{\ominus} - \sum_j n_j\mu_j^{\ominus} = -RT\left(\sum_k \ln {a_k}^{m_k} - \sum_j \ln {a_j}^{n_j}\right)$$

$$\Delta G^{\ominus} = -RT\ln K.$$

This relates the standard Gibbs free energy change, ΔG° to an equilibrium constant, K, the reaction quotient of activity values at equilibrium.

$$\Delta G^{\ominus} = \sum_k m_k\mu_k^{\ominus} - \sum_j n_j\mu_j^{\ominus}$$

$$\ln K = \sum_k \ln {a_k}^{m_k} - \sum_j \ln {a_j}^{n_j};\ K = \frac{\prod_k {a_k}^{m_k}}{\prod_j {a_j}^{n_j}} \equiv \frac{\{R\}^{\rho}\{S\}^{\sigma}\ldots}{\{A\}^{\alpha}\{B\}^{\beta}\ldots}$$

Temperature Dependence

The van 't Hoff equation:

$$\frac{d\ln K}{dT} = \frac{\Delta H^{\ominus}}{RT^2} \text{ or } \frac{d\ln K}{d(T^{-1})} = -\frac{\Delta H^{\ominus}}{R}$$

shows that when the reaction is exothermic (ΔH° is negative), then K decreases with increasing temperature, in accordance with Le Chatelier›s principle. It permits calculation of the reaction equilibrium constant at temperature T_2 if the reaction constant at T_1 is known and the standard reaction enthalpy can be assumed to be independent of temperature even though each standard enthalpy change is defined at a different temperature. However, this assumption is valid only for small temperature differences $T_2 - T_1$. In fact standard thermodynamic arguments can be used to show that

$$\left(\frac{\partial H}{\partial T}\right)_p = C_p$$

where C_p is the heat capacity at constant pressure. The equilibrium constant is related to the standard Gibbs energy change of reaction as

$$\Delta G^\ominus = -RT \ln K$$

where ΔG° is the standard Gibbs free energy change of reaction, R is the gas constant, and T the absolute temperature.

If the equilibrium constant has been determined and the standard reaction enthalpy has also been determined, by calorimetry, for example, this equation allows the standard entropy change for the reaction to be derived.

A More Complex Formulation

The calculation of K at a particular temperature from a known K at another given temperature can be approached as follows if standard thermodynamic properties are available. The effect of temperature on equilibrium constant is equivalent to the effect of temperature on Gibbs energy because:

$$\ln K = \frac{-\Delta_r G^\ominus}{RT}$$

where $\Delta_r G^\circ$ is the reaction standard Gibbs energy, which is the sum of the standard Gibbs energies of the reaction products minus the sum of standard Gibbs energies of reactants.

Here, the term "standard" denotes the ideal behaviour (i.e., an infinite dilution) and a hypothetical standard concentration (typically 1 mol/kg). It does not imply any particular temperature or pressure because, although contrary to IUPAC recommendation, it is more convenient when describing aqueous systems over wide temperature and pressure ranges.

The standard Gibbs energy (for each species or for the entire reaction) can be represented (from the basic definitions) as:

$$G^\ominus_{T_2} = G^\ominus_{T_1} - S^\ominus_{T_1}(T_2 - T_1) - T_2 \int_{T_1}^{T_2} \frac{C_p^\ominus}{T}\, dT + \int_{T_1}^{T_2} C_p^\ominus\, dT$$

In the above equation, the effect of temperature on Gibbs energy (and thus on the equilibrium constant) is ascribed entirely to heat capacity. To evaluate the integrals in this equation, the form of the dependence of heat capacity on temperature needs to be known.

Now, if one expresses the standard heat capacity $C_p^\ominus$, as a function of absolute temperature using correlations of the following forms:

- For pure substances (solids, gas, liquid):

$$C_p^\ominus = A + BT + CT^{-2}$$

- For ionic species at $T < 200$ °C:

$$C_p^\ominus = (4.186a + b\breve{S}^\ominus_{T_1})\frac{(T_2 - T_1)}{\ln\left(\frac{T_2}{T_1}\right)}$$

then the integrals can be evaluated and the following final form is obtained:

$$G^{\ominus}_{T_2} = G^{\ominus}_{T_1} + (C^{\ominus}_p - S^{\ominus}_{T_1})(T_2 - T_1) - T_2 \ln\left(\frac{T_2}{T_1}\right) C^{\ominus}_p$$

The constants A, B, C, a, b and the absolute entropy, $\check{S}^{\ominus}_{298\ K}$, required for evaluation of $C^{\ominus}_p(T)$, as well as the values of $G_{298\ K}$ and $S_{298\ K}$ for many species are tabulated in the literature.

Pressure Dependence

The pressure dependence of the equilibrium constant is usually weak in the range of pressures normally encountered in industry, and therefore, it is usually neglected in practice. This is true for condensed reactant/products (i.e., when reactants and products are solids or liquid) as well as gaseous ones.

For a gaseous-reaction example, one may consider the well-studied reaction of hydrogen with nitrogen to produce ammonia:

$$N_2 + 3\ H_2 \rightleftharpoons 2\ NH_3$$

If the pressure is increased by an addition of an inert gas, then neither the composition at equilibrium nor the equilibrium constant are appreciably affected (because the partial pressures remain constant, assuming an ideal-gas behaviour of all gases involved). However, the composition at equilibrium will depend appreciably on pressure when:

- the pressure is changed by compression of the gaseous reacting system, and
- the reaction results in the change of the number of moles of gas in the system.

In the example reaction above, the number of moles changes from 4 to 2, and an increase of pressure by system compression will result in appreciably more ammonia in the equilibrium mixture. In the general case of a gaseous reaction:

$$\alpha A + \beta B \rightleftharpoons \sigma S + \tau T$$

the change of mixture composition with pressure can be quantified using:

$$K_p = \frac{p_S{}^{\sigma} p_T{}^{\tau}}{p_A{}^{\alpha} p_B{}^{\beta}} = \frac{X_S{}^{\sigma} X_T{}^{\tau}}{X_A{}^{\alpha} X_B{}^{\beta}} P^{\sigma+\tau-\alpha-\beta} = K_X P^{\sigma+\tau-\alpha-\beta}$$

where p denote the partial pressures of the components, P is the total system pressure, X denote the mole fraction, K_p is the equilibrium constant expressed in terms of partial pressures and K_X is the equilibrium constant expressed in terms of mole fractions.

The above change in composition is in accordance with Le Chatelier's principle and does not involve any change of the equilibrium constant with the total system pressure. Indeed, for ideal-gas reactions K_p is independent of pressure.

In a condensed phase, the pressure dependence of the equilibrium constant is associated with the reaction molar volume.For reaction:

$$\alpha A + \beta B \rightleftharpoons \sigma S + \tau T$$

the reaction molar volume is:

$$\Delta\bar{V} = \sigma\bar{V}_S + \tau\bar{V}_T - \alpha\bar{V}_A - \beta\bar{V}_B$$

where V denotes a partial molar volume of a reactant or a product.

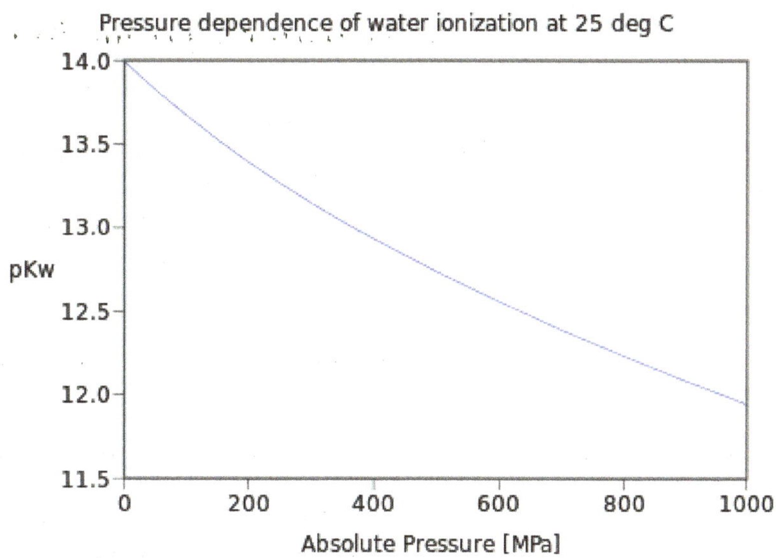

Pressure dependence of the water ionization constant at 25 °C. In general, ionization in aqueous solutions tends to increase with increasing pressure.

For the above reaction, one can expect the change of the reaction equilibrium constant (based either on mole-fraction or molal-concentration scale) with pressure at constant temperature to be:

$$\left(\frac{\partial \ln K_X}{\partial P}\right)_T = \frac{-\Delta\bar{V}}{RT}.$$

The matter is complicated as partial molar volume is itself dependent on pressure.

HENDERSON–HASSELBALCH EQUATION

In chemistry, the **Henderson–Hasselbalch equation** describes the derivation of pH as a measure of acidity (using pK_a, the negative log of the acid dissociation constant) in biological and chemical systems. The equation is also useful for estimating the pH of a buffer solution and finding the equilibrium pH in acid-base reactions (it is widely used to calculate the isoelectric point of proteins).

The equation is given by:

$$\mathrm{pH} = \mathrm{p}K_\mathrm{a} + \log_{10}\left(\frac{[\mathrm{A}^-]}{[\mathrm{HA}]}\right)$$

Here, [HA] is the molar concentration of the undissociated weak acid, [A$^-$] is the molar concentration (molarity, M) of this acid's conjugate base and pK_a is $-\log_{10} K_a$ where K_a is the acid dissociation constant, that is:

$$\mathrm{p}K_\mathrm{a} = -\log_{10}(K_\mathrm{a}) = -\log_{10}\left(\frac{[\mathrm{H_3O^+}][\mathrm{A}^-]}{[\mathrm{HA}]}\right)$$

for the non-specific Brønsted acid-base reaction: $\mathrm{HA} + \mathrm{H_2O} \rightleftharpoons \mathrm{A}^- + \mathrm{H_3O^+}$

In these equations, A$^-$ denotes the ionic form of the relevant acid. Bracketed quantities such as [base] and [acid] denote the molar concentration of the quantity enclosed.

For Bases

For the standard base equation:

$$\mathrm{B} + \mathrm{H^+} \rightleftharpoons \mathrm{BH^+}$$

A second form of the equation expressed in terms of K_b where K_b is the base dissociation constant:

$$\mathrm{p}K_\mathrm{b} = -\log_{10}(K_\mathrm{b}) = -\log_{10}\left(\frac{[\mathrm{OH}^-][\mathrm{HA}]}{[\mathrm{A}^-]}\right)$$

In analogy to the above equations, the following equation is valid:

$$\mathrm{pOH} = \mathrm{p}K_\mathrm{b} + \log_{10}\left(\frac{[\mathrm{BH^+}]}{[\mathrm{B}]}\right),$$

where BH$^+$ denotes the conjugate acid of the corresponding base B. Using the properties of these terms at 25 degrees Celsius one can synthesise an equation for pH of basic solutions in terms of pK_a and pH:

$$\mathrm{pH} = \mathrm{p}K_\mathrm{a} + \log_{10}\left(\frac{[\mathrm{B}]}{[\mathrm{BH^+}]}\right)$$

Derivation

The Henderson–Hasselbalch equation is derived from the acid dissociation constant equation by the following steps:

$$K_a = \frac{[H^+][A^-]}{[HA]}$$

Taking the log, to base ten, of both sides gives:

$$\log_{10} K_a = \log_{10}\left(\frac{[H^+][A^-]}{[HA]}\right)$$

Then, using the properties of logarithms:

$$\log_{10} K_a = \log_{10}[H^+] + \log_{10}\left(\frac{[A^-]}{[HA]}\right)$$

Identifying the left-hand side of this equation as $-pK_a$ and the $\log_{10}[H^+]$ as -pH:

$$-pK_a = -pH + \log_{10}\left(\frac{[A^-]}{[HA]}\right)$$

Adding pH and pK_a to both sides:

$$pH = pK_a + \log_{10}\left(\frac{[A^-]}{[HA]}\right)$$

The ratio $[A^-]/[HA]$ is unitless, and as such, other ratios with other units may be used. For example, the mole ratio of the components, n_{A^-}/n_{HA} or the fractional concentrations α_{A^-}/α_{HA} where $\alpha_{A^-} + \alpha_{HA} = 1$ will yield the same answer. Sometimes these other units are more convenient to use.

Limitations

There are some significant approximations implicit in the Henderson–Hasselbalch equation. The most significant is the assumption that the concentration of the acid and its conjugate base at equilibrium will remain the same as the formal concentration. This neglects the dissociation of the acid and the binding of H+ to the base. The dissociation of water and relative water concentration itself is neglected as well. These approximations will fail when dealing with relatively strong acids or bases (pKa more than a couple units away from 7), dilute or very concentrated solutions (less than 1 mM or greater than 1M), or heavily skewed acid/base ratios (more than 100 to 1). In high buffer dilutions, where the concentration of protons arising from water become equally or more prevalent than the buffer species themselves (at pH 7, this means buffer component concentrations of $<10^{-5}$ M formally, but practically much higher), the pKa of the 'buffer' system will tend towards neutrality.

Conventions

Note that it is conventional for basic compounds to quote a pK_a which refers to the pK_a of the conjugate acid rather than that of the basic compound itself, likewise for pK_b when discussing acids. Thus a low pK_a quoted for a basic compound means that the conjugate acid readily dissociates and therefore that the basic compound is likely to be neutral at physiological pH. This can be contrasted with a low pK_a for an acid, which means that the acid readily dissociates and the molecule is therefore likely to exist as its conjugate base at physiological pH. To interpret literature pK_a and pK_b values an inspection of the molecule to determine whether it contains acid or base functionality (or both) is important.

Estimating Blood pH

The Henderson–Hasselbalch equation can be applied to relate the pH of blood to constituents of the bicarbonate buffering system:

$$\mathrm{pH} = \mathrm{p}K_{\mathrm{a}\ \mathrm{H_2CO_3}} + \log_{10}\left(\frac{[\mathrm{HCO_3^-}]}{[\mathrm{H_2CO_3}]}\right)$$

where:

- pK_a H_2CO_3 is the cologarithm of the acid dissociation constant of carbonic acid. It is equal to 6.1.
- $[HCO_3-]$ is the concentration of bicarbonate in the blood
- $[H_2CO_3]$ is the concentration of carbonic acid in the blood

This is useful in arterial blood gas, but these usually state p_{CO2}, that is, the partial pressure of carbon dioxide, rather than H_2CO_3. However, these are related by the equation:

where:

- $[H_2CO_3]$ is the concentration of carbonic acid in the blood
- k_H CO_2 is the Henry's law constant for the solubility of carbon dioxide in blood. k_H CO_2 is approximately 0.0307 mmol/(L-torr)
- PCO_2 is the partial pressure of carbon dioxide in the blood

Taken together, the following equation can be used to relate the pH of blood to the concentration of bicarbonate and the partial pressure of carbon dioxide:

$$\mathrm{pH} = 6.1 + \log_{10}\left(\frac{[\mathrm{HCO_3^-}]}{0.0307 \times p_{\mathrm{CO_2}}}\right),$$

where:

- pH is the acidity in the blood
- $[HCO_3^-]$ is the concentration of bicarbonate in the blood
- p_{CO2} is the partial pressure of carbon dioxide in the arterial blood

MICHAELIS–MENTEN KINETICS

In biochemistry, 'Michaelis–Menten' kinetics is one of the best-known models of enzyme kinetics. It is named after German biochemist Leonor Michaelis and Canadian physician Maud Menten. The model takes the form of an equation describing the rate of enzymatic reactions, by relating reaction rate v to [S] the concentration of a substrate S. Its formula is given by

$$v = \frac{d[P]}{dt} = \frac{V_{\max}[S]}{K_{\mathrm{M}} + [S]}.$$

This equation is called the Michaelis–Menten equation. Here,v_{max} represents the maximum rate achieved by the system, at saturating substrate concentration. The Michaelis constant K_M is the substrate concentration at which the reaction rate is half of v_{max} Biochemical reactions involving a single substrate are often assumed to follow Michaelis–Menten kinetics, without regard to the model's underlying assumptions.

Model

In 1903, French physical chemist Victor Henri found that enzyme reactions were initiated by a bond (more generally, a binding interaction) between the enzyme and the substrate.His work was taken up by German biochemist Leonor Michaelis and Canadian physician Maud Menten, who investigated the kinetics of an enzymatic reaction mechanism, invertase, that catalyzes the hydrolysis of sucrose into glucose and fructose.In 1913, they proposed a mathematical model of the reaction.It involves an enzyme, E, binding to a substrate, S, to form a complex, ES, which in turn releases a product, P, regenerating the original enzyme. This may be represented schematically as

$$\mathrm{E} + \mathrm{S} \underset{k_r}{\overset{k_f}{\rightleftharpoons}} \mathrm{ES} \overset{k_{\mathrm{cat}}}{\longrightarrow} \mathrm{E} + \mathrm{P}$$

where k_f (forward rate), k_r (reverse rate), and k_{cat} (catalytic rate) denote the rate constants, the double arrows between S (substrate) and ES (enzyme-substrate complex) represent the fact that enzyme-substrate binding is a reversible process, and the single forward arrow represents the formation of P (product).

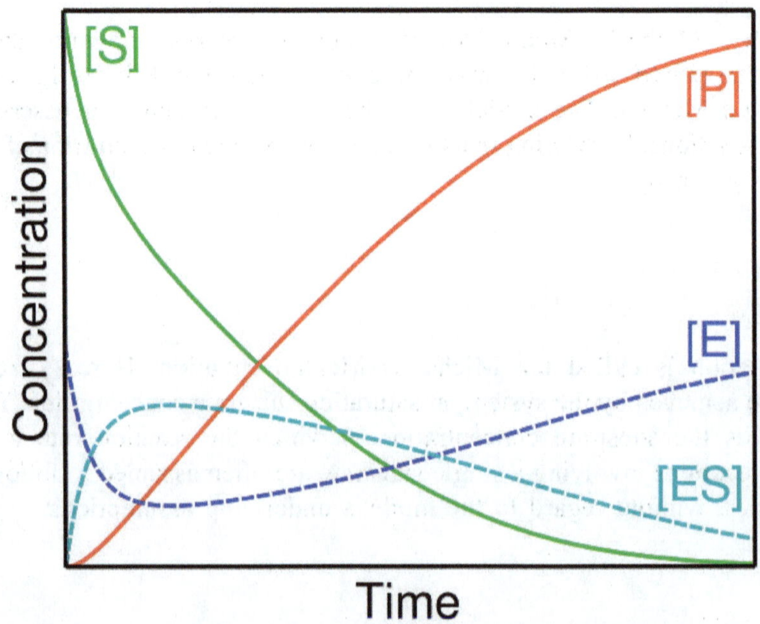

Change in concentrations over time for enzyme E, substrate S, complex ES and product P

Under certain assumptions – such as the enzyme concentration being much less than the substrate concentration – the rate of product formation is given by

$$v = \frac{d[\mathrm{P}]}{dt} = V_{\max}\frac{[\mathrm{S}]}{K_\mathrm{M} + [\mathrm{S}]} = k_\mathrm{cat}[\mathrm{E}]_0\frac{[\mathrm{S}]}{K_\mathrm{M} + [\mathrm{S}]}.$$

The reaction order depends on the relative size of the two terms in the denominator. At low substrate concentration $[S] \ll K_M$ so that $v = k_\mathrm{cat}[\mathrm{E}]_0\frac{[\mathrm{S}]}{K_\mathrm{M}}$. Under these conditions the reaction rate varies linearly with substrate concentration $[\mathrm{S}]$ (first-order kinetics). However at higher $[\mathrm{S}]$ with $[S] \gg K_M$, the reaction becomes independent of $[\mathrm{S}]$ (zero-order kinetics) and asymptotically approaches its maximum rate $V_\mathrm{max} = k_\mathrm{cat}[\mathrm{E}]_0$, where $[\mathrm{E}]_0$ is the initial enzyme concentration. This rate is attained when all enzyme is bound to substrate. k_cat the turnover number, is the maximum number of substrate molecules converted to product per enzyme molecule per second. Further addition of substrate does not increase the rate which is said to be saturated.

The Michaelis constant K_M is the $[\mathrm{S}]$ at which the reaction rate is at half-maximum, and is an inverse measure of the substrate's affinity for the enzyme—as a small K_M indicates high affinity, meaning that the rate will approach V_{MAX} with lower [S] than those reactions with a larger K_M The value of KM is dependent on both the enzyme and the substrate, as well as conditions such as temperature and pH.

The model is used in a variety of biochemical situations other than enzyme-substrate interaction, including antigen-antibody binding, DNA-DNA hybridization, and protein-protein interaction. It can be used to characterise a generic biochemical reaction, in the same way that the Langmuir equation can be used to model generic adsorption of biomolecular species. When an empirical equation of this form is applied to microbial growth, it is sometimes called a Monod equation.

Applications

Parameter values vary widely between enzymes:

Enzyme	K_M (M)	k_{cat} (1/s)	k_{cat}/K_M (1/M*s)
Chymotrypsin	1.5×10^{-2}	0.14	9.3
Pepsin	3.0×10^{-4}	0.50	1.7×10^{3}
Tyrosyl-tRNA synthetase	9.0×10^{-4}	7.6	8.4×10^{3}
Ribonuclease	7.9×10^{-3}	7.9×10^{2}	1.0×10^{5}
Carbonic anhydrase	2.6×10^{-2}	4.0×10^{5}	1.5×10^{7}
Fumarase	5.0×10^{-6}	8.0×10^{2}	1.6×10^{8}

The constant k_{cat}/K_M (catalytic efficiency) is a measure of how efficiently an enzyme converts a substrate into product. Diffusion limited enzymes, such as fumarase, work at the theoretical upper limit of 10^8 – 10^{10} /M*s, limited by diffusion of substrate into the active site.

Michaelis–Menten kinetics have also been applied to a variety of spheres outside of biochemical reactions, including alveolar clearance of dusts, the richness of species pools, clearance of blood alcohol, the photosynthesis-irradiance relationship, and bacterial phage infection.

Derivation

Applying the law of mass action, which states that the rate of a reaction is proportional to the product of the concentrations of the reactants (i.e.[E][S]), gives a system of four non-linear ordinary differential equations that define the rate of change of reactants with time t

$$\frac{d[E]}{dt} = -k_f[E][S] + k_r[ES] + k_{cat}[ES]$$
$$\frac{d[S]}{dt} = -k_f[E][S] + k_r[ES]$$
$$\frac{d[ES]}{dt} = k_f[E][S] - k_r[ES] - k_{cat}[ES]$$
$$\frac{d[P]}{dt} = k_{cat}[ES].$$

In this mechanism, the enzyme E is a catalyst, which only facilitates the reaction, so that its total concentration, free plus combined, $[E] + [ES] = [E]_0$ is a constant. This conservation law can also be observed by adding the first and third equations above.

Equilibrium Approximation

In their original analysis, Michaelis and Menten assumed that the substrate is in instantaneous chemical equilibrium with the complex, which implies

$$k_f[E][S] = k_r[ES].$$

From the enzyme conservation law, we obtain

$$[E] = [E]_0 - [ES].$$

Combining the two expressions above, gives us

$$k_f([E]_0 - [ES])[S] = k_r[ES].$$

Upon simplification, we get

$$[ES] = \frac{[E]_0[S]}{K_d + [S]}$$

where $K_d = k_r/k_f$ is the dissociation constant for the enzyme-substrate complex. Hence the velocity v of the reaction – the rate at which P is formed – is

$$v = \frac{d[P]}{dt} = \frac{V_{\max}[S]}{K_d + [S]}$$

where $V_{\max} = k_{\text{cat}}[E]_0$ is the maximum reaction velocity.

Quasi-Steady-State Approximation

An alternative analysis of the system was undertaken by British botanist G. E. Briggs and British geneticist J. B. S. Haldane in 1925. They assumed that the concentration of the intermediate complex does not change on the time-scale of product formation – known as the quasi-steady-state assumption or pseudo-steady-state-hypothesis. Mathematically, this assumption means

$k_f[E][S] = k_r[ES] + k_{\mathrm{cat}}[ES]$. Combining this relationship with the enzyme conservation law, the concentration of the complex is

$$[ES] = \frac{[E]_0[S]}{K_{\mathrm{M}} + [S]}$$

where

$$K_{\mathrm{M}} = \frac{k_r + k_{\mathrm{cat}}}{k_f}$$

is known as the Michaelis constant, where k_r, k_{cat}, and k_f are, respectively, the constants for substrate unbinding, conversion to product, and binding to the enzyme. Hence the velocity v of the reaction is

$$v = \frac{d[P]}{dt} = \frac{V_{\max}[S]}{K_{\mathrm{M}} + [S]}.$$

Assumptions and Limitations

The first step in the derivation applies the law of mass action, which is reliant on free diffusion. However, in the environment of a living cell where there is a high concentration of proteins, the cytoplasm often behaves more like a gel than a liquid, limiting molecular movements and altering reaction rates. Although the law of mass action can be valid in heterogeneous environments, it is more appropriate to model the cytoplasm as a fractal, in order to capture its limited-mobility kinetics.

The resulting reaction rates predicted by the two approaches are similar, with the only difference being that the equilibrium approximation defines the constant as k_d whilst the quasi-steady-state approximation uses KM However, each approach is founded upon a different assumption. The Michaelis–Menten equilibrium analysis is valid if the substrate reaches equilibrium on a much faster time-scale than the product is formed or, more precisely, that

$$\epsilon_d = \frac{k_{\mathrm{cat}}}{k_r} \ll 1.$$

By contrast, the Briggs–Haldane quasi-steady-state analysis is valid if

$$\epsilon_m = \frac{[\mathrm{E}]_0}{[\mathrm{S}]_0 + K_\mathrm{M}} \ll 1.$$

Thus it holds if the enzyme concentration is much less than the substrate concentration. Even if this is not satisfied, the approximation is valid if K_M is large.

In both the Michaelis–Menten and Briggs–Haldane analyses, the quality of the approximation improves as ϵ decreases. However, in model building, Michaelis–Menten kinetics are often invoked without regard to the underlying assumptions.

It is also important to remember that, while irreversibility is a necessary simplification in order to yield a tractable analytic solution, in the general case product formation is not in fact irreversible. The enzyme reaction is more correctly described as

$$\mathrm{E} + \mathrm{S} \underset{k_{r_1}}{\overset{k_{f_1}}{\rightleftharpoons}} \mathrm{ES} \underset{k_{r_2}}{\overset{k_{f_2}}{\rightleftharpoons}} \mathrm{E} + \mathrm{P}.$$

In general, the assumption of irreversibility is a good one in situations where one of the below is true:

1. The concentration of substrate(s) is very much larger than the concentration of products:

$$[\mathrm{S}] \gg [\mathrm{P}].$$

This is true under standard in vitro assay conditions, and is true for many in vivo biological reactions, particularly where the product is continually removed by a subsequent reaction.

2. The energy released in the reaction is very large, that is

$$\Delta G \ll 0.$$

In situations where neither of these two conditions hold (that is, the reaction is low energy and a substantial pool of product(s) exists), the Michaelis–Menten equation breaks down, and more complex modelling approaches explicitly taking the forward and reverse reactions into account must be taken to understand the enzyme biology.

Determination of Constants

The typical method for determining the constants V_max and K_M involves running a series of enzyme assays at varying substrate concentrations $[S]$, and measuring the initial reaction rate v_0. Initial' here is taken to mean that the reaction rate is measured after a relatively short time period, during which it is assumed that the enzyme-substrate complex has formed, but that the substrate concentration held approximately constant, and so the equilibrium or quasi-steady-state approximation remain valid. By plotting reaction rate against concentration, and using nonlinear regression of the Michaelis–Menten equation, the parameters may be obtained.

Before computing facilities to perform nonlinear regression became available, graphical methods involving linearisation of the equation were used. A number of these were proposed, including the Eadie–Hofstee diagram, Hanes–Woolf plot and Lineweaver–Burk plot; of these, the Hanes–Woolf plot is the most accurate. However, while useful for visualization, all three methods distort the error structure of the data and are inferior to nonlinear regression. Nonetheless, their use can still be found in modern literature.

In 1997 Santiago Schnell and Claudio Mendoza suggested a closed form solution for the time course kinetics analysis of the Michaelis–Menten kinetics based on the solution of the Lambert W function. Namely:

$$\frac{[S]}{K_M} = W(F(t))$$

where W is the Lambert W function and

$$F(t) = \frac{[S]_0}{K_M} \exp\left(\frac{[S]_0}{K_M} - \frac{V_{max}}{K_M} t\right).$$

The above equation has been used to estimate V_{max} and K_M from time course data.

Role of Substrate Unbinding

The Michaelis-Menten equation has been used to predict the rate of product formation in enzymatic reactions for more than a century. Specifically, it states that the rate of an enzymatic reaction will increase as substrate concentration increases, and that increased unbinding of enzyme-substrate complexes will decrease the reaction rate. While the first prediction is well established, the second has never been tested experimentally. To determine whether an increased rate of unbinding does in fact decrease the reaction rate, Shlomi Reuveni et al. mathematically analyzed the effect of enzyme-substrate unbinding on enzymatic reactions at the single-molecule level. According to the study, unbinding of an enzyme from a substrate can reduce the rate of product formation under some conditions, but may also have the opposite effect. As substrate concentrations increase, a tipping point can be reached where an increase in the unbinding rate results in an increase, rather than a decrease, of the reaction rate. The results indicate that enzymatic reactions can behave in ways that violate the classical Michaelis-Menten equation, and that the role of unbinding in enzymatic catalysis still remains to be determined experimentally.

REFERENCES

1. Atkins, Peter; De Paula, Julio (2006). Atkins' Physical Chemistry (8th ed.). W. H. Freeman. pp. 200–202. ISBN 0-7167-8759-8.
2. Atkins, Peter W.; Jones, Loretta. Chemical Principles: The Quest for Insight (2nd ed.). ISBN 0-7167-9903-0.

3. IUPAC, Compendium of Chemical Terminology, 2nd ed. (the "Gold Book") (1997). Online corrected version: (2006–) "chemical equilibrium".
4. Brady, James E. Chemistry: Matter and Its Changes (4th ed.). Fred Senese. ISBN 0-471-21517-1.
5. Schultz, Mary Jane (1999). "Why Equilibrium? Understanding Entropy of Mixing". Journal of Chemical Education. 76 (10): 1391. Bibcode:1999JChEd..76.1391S. doi:10.1021/ed076p1391.
6. Clugston, Michael J. (1990). "A mathematical verification of the second law of thermodynamics from the entropy of mixing". Journal of Chemical Education. 67 (3): 203. Bibcode:1990JChEd..67Q.203C. doi:10.1021/ed067p203.
7. Davies, C. W. (1962). Ion Association. Butterworths.
8. Grenthe, I.; Wanner, H. "Guidelines for the extrapolation to zero ionic strength" (PDF).
9. Rossotti, F. J. C.; Rossotti, H. (1961). The Determination of Stability Constants. McGraw-Hill.
10. Concise Encyclopedia Chemistry. 1994. ISBN 0-89925-457-8.
11. Beck, M. T.; Nagypál, I. (1990). Chemistry of Complex Equilibria (2nd ed.). Budapest: Akadémiai Kaidó.
12. "Computer Program for Calculation of Complex Chemical Equilibrium Compositions and Applications" (PDF). NASA Reference publication 1311. NASA. 1994.
13. Smith, W. R.; Missen, R. W. (1991). Chemical Reaction Equilibrium Analysis: Theory and Algorithms (Reprinted ed.). Malabar, FL: Krieger Publishing.
14. "Mathtrek Systems".
15. The diagram was created with the program HySS
16. Larsen, D. "Henderson-Hasselbalch Approximation". Chemwiki. University of California. Retrieved 27 March 2014.
17. Henderson Hasselbalch Equation: Derivation of pKa and pKb
18. "Acid-Base Tutorial - History". www.acid-base.com. Retrieved 2016-04-04.
19. page 556, section "Estimating plasma pH" in: Bray, John J. (1999). Lecture notes on human physiolog. Malden, Mass.: Blackwell Science. ISBN 978-0-86542-775-4.
20. Rossotti, F. J. C.; Rossotti, H. (1961). The Determination of Stability Constants. McGraw-Hill. p. 5.
21. IUPAC (2011). Green Book (PDF) (3rd ed.). p. 58.
22. Hague, David N.; Moreton, Anthony D. (1994). "Protonation sequence of linear aliphatic polyamines by 13C NMR spectroscopy". J. Chem. Soc., Perkin Trans. 2 (2): 265–70. doi:10.1039/P29940000265.
23. Borkovec, Michal; Koper, Ger J. M. (2000). "A Cluster Expansion Method for the Complete Resolution of Microscopic Ionization Equilibria from NMR Titrations". Anal. Chem. 72 (14): 3272–9. doi:10.1021/ac991494p. PMID 10939399.

24. Baes, C. F.; Mesmer, R. E. (1976). The Hydrolysis of Cations. Wiley.
25. Schwarzenbach, G.; Flaschka, H. (1969). Complexometric titrations. Methuen.
26. Rossotti, F. J. C.; Rossotti, H. (1961). "Chapter 2: Activity and Concentration Quotients". The Determination of Stability Constants. McGraw–Hill.
27. Butler, J. N. (1998). Ionic Equilibrium. John Wiley and Sons.
28. "Project: Ionic Strength Corrections for Stability Constants". International Union of Pure and Applied Chemistry. Archived from the original on 29 October 2008. Retrieved 2008-11-23.
29. Denbigh, K. (1981). "Chapter 4". The principles of chemical equilibrium (4th ed.). Cambridge: Cambridge University Press. ISBN 0-521-28150-4.
30. Majer, V.; Sedelbauer, J.; Wood (2004). "Calculations of standard thermodynamic properties of aqueous electrolytes and nonelectrolytes". In Palmer, D. A.; Fernandez-Prini, R.; Harvey, A. Aqueous Systems at Elevated Temperatures and Pressures: Physical Chemistry of Water, Steam and Hydrothermal Solutions. Elsevier.
31. Roberge, P. R. (2000). "Appendix F". Handbook of Corrosion Engineering. McGraw-Hill.
32. Atkins, P. W. (1978). Physical Chemistry, (6th ed.). Oxford University Press. p. 210.
33. Van Eldik, R.; Asano, T.; Le Noble, W. J. (1989). "Activation and reaction volumes in solution. 2". Chem. Rev. 89 (3): 549–688. doi:10.1021/cr00093a005.
34. http://www.worthington-biochem.com/introbiochem/substrateconc.html
35. Henri, Victor (1903). Lois Générales de l'Action des Diastases. Paris: Hermann. Google books (US only)
36. "Victor Henri". Whonamedit?. Retrieved 24 May 2011.
37. Michaelis, L.; Menten, M.L. (1913). "Die Kinetik der Invertinwirkung". Biochem Z. 49: 333–369 (recent translation, and an older partial translation)
38. Chen, W.W.; Neipel, M.; Sorger, P.K. (2010). "Classic and contemporary approaches to modeling biochemical reactions". Genes Dev. 24 (17): 1861–1875. doi:10.1101/gad.1945410. PMC 2932968 . PMID 20810646.
39. Laidler K.J. and Meiser J.H. Physical Chemistry (Benjamin/Cummings 1982) p.430 ISBN 0-8053-5682-7
40. Lehninger, A.L.; Nelson, D.L.; Cox, M.M. (2005). Lehninger principles of biochemistry. New York: W.H. Freeman. ISBN 978-0-7167-4339-2.
41. Chakraborty, S. (23 Dec 2009). Microfluidics and Microfabrication (1 ed.). Springer. ISBN 978-1-4419-1542-9.
42. Mathews, C.K.; van Holde, K.E.; Ahern, K.G. (10 Dec 1999). Biochemistry (3 ed.). Prentice Hall. ISBN 978-0-8053-3066-3.
43. Stroppolo, M.E.; Falconi, M.; Caccuri, A.M.; Desideri, A. (Sep 2001). "Superefficient enzymes". Cell Mol Life Sci. 58 (10): 1451–60. doi:10.1007/PL00000788. PMID 11693526.

44. Yu, R.C.; Rappaport, S.M. (1997). "A lung retention model based on Michaelis–Menten-like kinetics". Environ Health Perspect. 105 (5): 496–503. doi:10.1289/ehp.97105496. PMC 1469867 . PMID 9222134.
45. Keating, K.A.; Quinn, J.F. (1998). "Estimating species richness: the Michaelis–Menten model revisited". Oikos. 81 (2): 411–416. doi:10.2307/3547060. JSTOR 3547060.
46. Jones, A.W. (2010). "Evidence-based survey of the elimination rates of ethanol from blood with applications in forensic casework". Forensic Sci Int. 200 (1–3): 1–20. doi:10.1016/j.forsciint.2010.02.021. PMID 20304569.
47. Abedon, S.T. (2009). "Kinetics of phage-mediated biocontrol of bacteria". Foodborne Pathog Dis. 6 (7): 807–15. doi:10.1089/fpd.2008.0242. PMID 19459758.
48. Murray, J.D. (2002). Mathematical Biology: I. An Introduction (3 ed.). Springer. ISBN 978-0-387-95223-9.
49. Keener, J.; Sneyd, J. (2008). Mathematical Physiology: I: Cellular Physiology (2 ed.). Springer. ISBN 978-0-387-75846-6.
50. Briggs, G.E.; Haldane, J.B.S. (1925). "A note on the kinematics of enzyme action". Biochem J. 19 (2): 338–339. doi:10.1042/bj0190338. PMC 1259181 . PMID 16743508.
51. Zhou, H.X.; Rivas, G.; Minton, A.P. (2008). "Macromolecular crowding and confinement: biochemical, biophysical, and potential physiological consequences". Annu Rev Biophys. 37 (1): 375–97. doi:10.1146/annurev.biophys.37.032807.125817. PMC 2826134 . PMID 18573087.
52. Grima, R.; Schnell, S. (Oct 2006). "A systematic investigation of the rate laws valid in intracellular environments". Biophys Chem. 124 (1): 1–10. doi:10.1016/j.bpc.2006.04.019. PMID 16781049.
53. Schnell, S.; Turner, T.E. (2004). "Reaction kinetics in intracellular environments with macromolecular crowding: simulations and rate laws". Prog Biophys Mol Biol. 85 (2–3): 235–60. doi:10.1016/j.pbiomolbio.2004.01.012. PMID 15142746.
54. ^ Jump up to:a b Segel, L.A.; Slemrod, M. (1989). "The quasi-steady-state assumption: A case study in perturbation". Thermochim Acta. 31 (3): 446–477. doi:10.1137/1031091.
55. Leskovac, V. (2003). Comprehensive enzyme kinetics. New York: Kluwer Academic/Plenum Pub. ISBN 978-0-306-46712-7.
56. Greco, W.R.; Hakala, M.T. (1979). "Evaluation of methods for estimating the dissociation constant of tight binding enzyme inhibitors,". J Biol Chem. 254 (23): 12104–12109. PMID 500698.
57. Hayakawa, K.; Guo, L.; Terentyeva, E.A.; Li, X.K.; Kimura, H.; Hirano, M.; Yoshikawa, K.; Nagamine, T.; et al. (2006). "Determination of specific activities and kinetic constants of biotinidase and lipoamidase in LEW rat and Lactobacillus casei (Shirota)". J Chromatogr B Analyt Technol Biomed Life Sci. 844 (2): 240–50. doi:10.1016/j.jchromb.2006.07.006. PMID 16876490.

58. Schnell, S.; Mendoza, C. (1997). "A closed form solution for time-dependent enzyme kinetics". Journal of Theoretical Biology. 187 (2): 207–212. doi:10.1006/jtbi.1997.0425.

59. Goudar, C. T.; Sonnad, J. R.; Duggleby, R. G. (1999). "Parameter estimation using a direct solution of the integrated Michaelis–Menten equation". Biochimica et Biophysica Acta (BBA) - Protein Structure and Molecular Enzymology. 1429 (2): 377–383. doi:10.1016/s0167-4838(98)00247-7. PMID 9989222.

60. Goudar, C. T.; Harris, S. K.; McInerney, M. J.; Suflita, J. M. (2004). "Progress curve analysis for enzyme and microbial kinetic reactions using explicit solutions based on the Lambert W function". Journal of Microbiological Methods. 59 (3): 317–326. doi:10.1016/j.mimet.2004.06.013. PMID 15488275.

61. Reuveni, Shlomi; Urbakh, Michael; Klafter, Joseph (2014). "Role of Substrate Unbinding in Michaelis-Menten Enzymatic Reactions". Proceedings of the National Academy of Sciences. 111 (12): 4391–4396. Bibcode:2014PNAS..111.4391R. doi:10.1073/pnas.1318122111.

CHAPTER 8

Chemical Composition

Chapter Contents

- Introduction
- Mole Fraction
- Volume Fraction
- Mass Fraction (Chemistry)
- Molality
- Molar Concentration
- Equivalent Concentration

The chemical composition of saliva is 98 percent water and 2 percent other substances such as electrolytes or minerals (sodium and potassium), mucus (mucopolysaccharides and glycoproteins), antibacterial compounds (thiocyanate and hydrogen peroxide) and enzymes. Enzymes, such as amylase, aid in the breakdown of carbohydrates and fats.

INTRODUCTION

Chemical composition, also known as chemical makeup or simply, composition, is a concept in chemistry that has different, but similar, meanings if referred to a single pure substance or a mixture.

Composition of a Substance

The chemical composition of a pure substance it corresponds to the relative amounts of the elements that constitute the substance itself. It can be expressed by the empirical formula.

For example the formula for water is H_2O: this means that each molecule it is constituted by 2 atoms of hydrogen (H) and 1 atom of oxygen (O) .

Composition of a Mixture

The chemical composition of a mixture can be defined as the distribution of the single substances that constitute the mixture, called "components". In other words, it is defined giving the concentration of each component.

Because there are different ways to define the concentration of a component, as a consequence there are also different ways to define the composition of a mixture. For example it can be expressed as

- molar fraction
- volume fraction
- mass fraction
- molality
- molarity
- normality.

MOLE FRACTION

In chemistry, the mole fraction or molar fraction (x_i) is defined as the amount of a constituent (expressed in moles), n_i, divided by the total amount of all constituents in a mixture, n_{tot}:

$$x_i = \frac{n_i}{n_{\mathrm{tot}}}$$

The sum of all the mole fractions is equal to 1:

$$\sum_{i=1}^{N} n_i = n_{\mathrm{tot}}; \sum_{i=1}^{N} x_i = 1$$

The same concept expressed with a denominator of 100 is the mole percent or molar percentage or molar proportion (mol%).

The mole fraction is also called the amount fraction. It is identical to the number fraction, which is defined as the number of molecules of a constituent Ni divided by the total number of all molecules Ntot. The mole fraction is sometimes denoted by the lowercase Greek letter χ (chi) instead of a Roman x. For mixtures of gases, IUPAC recommends the letter y.

The National Institute of Standards and Technology of the United States prefers the term amount-of-substance fraction over mole fraction because it does not contain the name of the unit mole.

Whereas mole fraction is a ratio of moles to moles, molar concentration is a quotient of moles to volume.

The mole fraction is one way of expressing the composition of a mixture with a dimensionless quantity; mass fraction (percentage by weight, wt%) and volume fraction (percentage by volume, vol%) are others.

Properties

Mole fraction is used very frequently in the construction of phase diagrams. It has a number of advantages:

- it is not temperature dependent (such as molar concentration) and does not require knowledge of the densities of the phase(s) involved
- a mixture of known mole fraction can be prepared by weighing off the appropriate masses of the constituents
- the measure is *symmetric*: in the mole fractions $x = 0.1$ and $x = 0.9$, the roles of 'solvent' and 'solute' are reversed.
- In a mixture of ideal gases, the mole fraction can be expressed as the ratio of partial pressure to total pressure of the mixture

Related Quantities

Mass Fraction

The mass fraction wi can be calculated using the formula

$$w_i = x_i \cdot \frac{M_i}{M}$$

where M_i is the molar mass of the component i and M is the average molar mass of the mixture.

Replacing the expression of the molar mass:

$$w_i = x_i \cdot \frac{M_i}{\sum_i x_i M_i}$$

Molar Mixing Ratio

The mixing of two pure components can be expressed introducing the amount or molar mixing ratio of them $r_n = \frac{n_2}{n_1}$. Then the mole fractions of the components will be:

$$x_1 = \frac{1}{1 + r_n}$$

$$x_2 = \frac{r_n}{1 + r_n}$$

Mole Percentage

Multiplying mole fraction by 100 gives the mole percentage, also referred as amount/amount percent (abbreviated as n/n%).

Mass Concentration

The conversion to and from mass concentration ρi is given by:

$$x_i = \frac{\rho_i}{\rho} \cdot \frac{M}{M_i}$$

where M is the average molar mass of the mixture.

$$\rho_i = x_i \rho \cdot \frac{M_i}{M}$$

Molar Concentration

The conversion to molar concentration ci is given by:

$$c_i = \frac{x_i \cdot \rho}{M} = x_i c$$

or

$$c_i = \frac{x_i \cdot \rho}{\sum_i x_i M_i}$$

where M is the average molar mass of the solution, c is the total molar concentration and ρ is the density of the solution.

Mass and Molar Mass

The mole fraction can be calculated from the masses mi and molar masses Mi of the components:

$$x_i = \frac{\frac{m_i}{M_i}}{\sum_i \frac{m_i}{M_i}}$$

Spatial Variation and Gradient

In a spatially non-uniform mixture, the mole fraction gradient triggers the phenomenon of diffusion.

VOLUME FRACTION

In chemistry, the volume fraction φi is defined as the volume of a constituent Vi divided by the volume of all constituents of the mixture V prior to mixing:

$$\phi_i = \frac{V_i}{\sum_j V_j}$$

Being dimensionless, its unit is 1; it is expressed as a number, e.g., 0.18. It is the same concept as volume percent (vol%) except that the latter is expressed with a denominator of 100, e.g., 18%.

The volume fraction coincides with the volume concentration in ideal solutions where the volumes of the constituents are additive (the volume of the solution is equal to the sum of the volumes of its ingredients).

The sum of all volume fractions of a mixture is equal to 1:

$$\sum_{i=1}^{N} V_i = V; \qquad \sum_{i=1}^{N} \phi_i = 1$$

The volume fraction (percentage by volume, vol%) is one way of expressing the composition of a mixture with a dimensionless quantity; mass fraction (percentage by weight, wt%) and mole fraction (percentage by moles, mol%) are others.

Volume Percent

Volume percent is a common expression of a solution's concentration. It is the volume fraction expressed with a denominator of 100, and thus is defined as:

$$\text{volume percent} = \frac{\text{volume of solute}}{\text{volume of solution}} \times 100$$

Volume percent is usually used when the solution is made by mixing two fluids, such as liquids or gases. However, percentages are only additive for ideal gases.

The percentage by volume (vol%) is one way of expressing the composition of a mixture with a dimensionless quantity; mass fraction (percentage by weight, wt%) and mole fraction (percentage by moles, mol%) are others.

In the case of a mixture of ethanol and water, which are miscible in all proportions, the designation of solvent and solute is arbitrary. The volume of such a mixture is slightly less than the sum of the volumes of the components. Thus, by the above definition, the term "40% alcohol by volume" refers to a mixture of 40 volume units of ethanol with enough water to make a final volume of 100 units, rather than a mixture of 40 units of ethanol with 60 units of water.

MASS FRACTION (CHEMISTRY)

In chemistry, the mass fraction w_i is the ratio of one substance with mass m_i to the mass of the total mixture m_{tot}, defined as

$$w_i = \frac{m_i}{m_{\text{tot}}}$$

The sum of all the mass fractions is equal to 1:

$$\sum_{i=1}^{N} m_i = m_{\text{tot}}; \sum_{i=1}^{N} w_i = 1$$

Mass fraction can also be expressed, with a denominator of 100, as percentage by mass (frequently, though erroneously, called percentage by weight, abbreviated wt%). It is one way of expressing the composition of a mixture in a dimensionless size; mole fraction (percentage by moles, mol%) and volume fraction (percentage by volume, vol%) are others.

For elemental analysis, mass fraction (or "mass percent composition") can also refer to the ratio of the mass of one element to the total mass of a compound. It can be calculated for any compound using its empirical formula or its chemical formula

Terminology

"Percent concentration" does not refer to this quantity. This improper name persists, especially in elementary textbooks. In biology, the unit "%" is sometimes (incorrectly) used to denote mass concentration, also called "mass/volume percentage." A solution with 1 g of solute dissolved in a final volume of 100 mL of solution would be labeled as "1 %" or "1 % m/v" (mass/volume). This is incorrect because the unit "%" can only be used for dimensionless quantities. Instead, the concentration should simply be given in units of g/mL. "Percent solution" or "percentage solution" are thus terms best reserved for "mass percent solutions" (m/m = m% = mass solute/mass total solution after mixing), or "volume percent solutions" (v/v = v% = volume solute per volume of total solution after mixing). The very ambiguous terms "percent solution" and "percentage solutions" with no other qualifiers continue to occasionally be encountered.

In thermal engineering vapor quality is used for the mass fraction of vapor in the steam.

In alloys, especially those of noble metals, the term fineness is used for the mass fraction of the noble metal in the alloy.

Properties

The mass fraction is independent of temperature.

Related Quantities

Mixing Ratio

The mixing of two pure components can be expressed introducing the (mass) mixing ratio of them

$r_m = \frac{m_2}{m_1}$. Then the mass fractions of the components will be:

$$w_1 = \frac{1}{1 + r_m}$$
$$w_2 = \frac{r_m}{1 + r_m}$$

Mass Concentration

The mass fraction of a component in a solution is the ratio of the mass concentration of that component ρi (density of that component in the mixture) to the density of solution ρ.

$$w_i = \frac{\rho_i}{\rho}$$

Molar Concentration

The relation to molar concentration is like that from above substituting the relation between mass and molar concentration.

$$w_i = \frac{\rho_i}{\rho} = \frac{c_i M_i}{\rho}$$

Mass Percentage

Multiplying mass fraction by 100 gives the mass percentage. It is sometimes called weight percent (wt%) or weight-weight percentage.

Mole Fraction

The mole fraction x_i can be calculated using the formula

$$x_i = w_i \cdot \frac{M}{M_i}$$

where M_i is the molar mass of the component *i* and *M* is the average molar mass of the mixture.

Replacing the expression of the molar mass-products:

$$x_i = \frac{\frac{w_i}{M_i}}{\sum_i \frac{w_i}{M_i}}$$

MOLALITY

Molality, also called molal concentration, is a measure of the concentration of a solute in a solution in terms of amount of substance in a specified amount of mass of the solvent. This contrasts with the definition of molarity which is based on a specified volume of solution. A commonly used unit for molality in chemistry is mol/kg. A solution of concentration 1 mol/kg is also sometimes denoted as 1 molal.

Definition

The molality (b), of a solution is defined as the amount of substance (in moles) of solute, n_{solute}, divided by the mass (in kg) of the solvent, $m_{solvent}$:

$$b = \frac{n_{\text{solute}}}{m_{\text{solvent}}}$$

In the cases of solutions with more than one solvent, molality can be defined for the mixed solvent considered as a pure pseudo-solvent. Instead of mole solut per kilogram solvent as in the binary case, units are defined as mole solut per kilogram mixed solvent.

Origin

The term molality is formed in analogy to molarity which is the molar concentration of a solution. The earliest known use of the intensive property molality and of its adjectival unit, the now-deprecated molal, appears to have been published by G. N. Lewis and M. Randall in the 1923 publication of Thermodynamics and the Free Energies of Chemical Substances. Though the two terms are subject to being confused with one another, the molality and molarity of a weak aqueous solution are nearly the same, as one kilogram of water (solvent) occupies the volume of 1 liter at room temperature and a small amount of solute has little effect on the volume.

Unit

The SI unit for molality is moles per kilogram. A solution with a molality of 3 mol/kg is often described as "3 molal" or "3 m". However, following the SI system of units, the National Institute of Standards and Technology, the United States authority on measurement, considers the term "molal" and the unit symbol "m" to be obsolete, and suggests mol/kg or a related unit of the SI. This recommendation has not been universally implemented in academia yet.

Usage Considerations

Advantages

The primary advantage of using molality as a measure of concentration is that molality only depends on the masses of solute and solvent, which are unaffected by variations in temperature and pressure. In contrast, solutions prepared volumetrically (e.g. molar concentration or mass concentration) are likely to change as temperature and pressure change. In many applications, this is a significant advantage because the mass, or the amount, of a substance is often more important than its volume (e.g. in a limiting reagent problem).

Another advantage of molality is the fact that the molality of one solute in a solution is independent of the presence or absence of other solutes.

Problem Areas

Unlike all the other compositional properties listed in "Relation" section (below), molality depends on the choice of the substance to be called "solvent" in an arbitrary mixture. If there is only one pure liquid substance in a mixture, the choice is clear, but not all solutions are this clear-cut: in an alcohol-water solution, either one could be called the solvent; in an alloy, or solid solution, there is no clear choice and all constituents may be treated alike. In such situations, mass or mole fraction is the preferred compositional specification.

Relation to other Compositional Properties

In what follows, the solvent may be given the same treatment as the other constituents of the solution, such that the molality of the solvent of an n-solute solution, say b_0, is found to be nothing more than the reciprocal of its molar mass, M_0:

$$b_0 = \frac{n_0}{n_0 M_0} = \frac{1}{M_0}.$$

Mass Fraction

The conversions to and from the mass fraction, w, of the solute in a single-solute solution are

$$w = \frac{1}{1 + \frac{1}{bM}}, \quad b = \frac{w}{(1-w)M},$$

where b is the molality and M is the molar mass of the solute.

More generally, for an n-solute/one-solvent solution, letting b_i and w_i be, respectively, the molality and mass fraction of the i-th solute,

$$w_i = w_0 b_i M_i, \quad b_i = \frac{w_i}{w_0 M_i},$$

where M_i is the molar mass of the ith solute, and w_0 is the mass fraction of the solvent, which is expressible both as a function of the molalities as well as a function of the other mass fractions,

$$w_0 = \frac{1}{1 + \sum_{j=1}^{n} b_j M_j} = 1 - \sum_{j=1}^{n} w_j.$$

Mole Fraction

The conversions to and from the mole fraction, x, of the solute in a single-solute solution are

$$x = \frac{1}{1 + \frac{1}{M_0 b}}, \quad b = \frac{x}{M_0(1-x)},$$

where M_0 is the molar mass of the solvent.

More generally, for an n-solute/one-solvent solution, letting x_i be the mole fraction of the ith solute,

$$x_i = x_0 M_0 b_i, \quad b_i = \frac{b_0 x_i}{x_0},$$

where x_0 is the mole fraction of the solvent, expressible both as a function of the molalities as well as a function of the other mole fractions:

$$x_0 = \frac{1}{1 + M_0 \sum_{j=1}^{n} b_j} = 1 - \sum_{j=1}^{n} x_j.$$

Molar Concentration (Molarity)

The conversions to and from the molar concentration, c, for one-solute solutions are

$$c = \frac{\rho b}{1 + bM}, \quad b = \frac{c}{\rho - cM};$$

where ρ is the mass density of the solution, b is the molality, and M is the molar mass of the solute.

For solutions with n solutes, the conversions are

$$c_i = c_0 M_0 b_i, \quad b_i = \frac{b_0 c_i}{c_0},$$

where the molar concentration of the solvent c_0 is expressible both as a function of the molalities as well as a function of the molarities:

$$c_0 = \frac{\rho b_0}{1 + \sum_{j=1}^{n} b_j M_j} = \frac{\rho - \sum_{j=1}^{n} c_i M_i}{M_0}.$$

Mass Concentration

The conversions to and from the mass concentration, ρ_{solute}, of a single-solute solution are

$$\rho_{\text{solute}} = \frac{\rho b M}{1 + bM}, \quad b = \frac{\rho_{\text{solute}}}{M(\rho - \rho_{\text{solute}})},$$

where ρ is the mass density of the solution, b is the molality, and M is the molar mass of the solute.

For the general n-solute solution, the mass concentration of the ith solute, ρ_i, is related to its molality, b_i, as follows:

$$\rho_i = \rho_0 b_i M_i, \quad b_i = \frac{\rho_i}{\rho_0 M_i},$$

where the mass concentration of the solvent, ρ_0, is expressible both as a function of the molalities as well as a function of the mass concentrations:

$$\rho_0 = \frac{\rho}{1 + \sum_{j=1}^{n} b_j M_j} = \rho - \sum_{j=1}^{n} \rho_i.$$

Equal Ratios

Alternatively, we may use just the last two equations given for the compositional property of the solvent in each of the preceding sections, together with the relationships given below, to derive the remainder of properties in that set:

$$\frac{b_i}{b_j} = \frac{x_i}{x_j} = \frac{c_i}{c_j} = \frac{\rho_i M_j}{\rho_j M_i} = \frac{w_i M_j}{w_j M_i},$$

where i and j are subscripts representing *all* the constituents, the n solutes plus the solvent.

Example of Conversion

An acid mixture consists of 0.76/0.04/0.20 mass fractions of 70% HNO_3 / 49% HF / H_2O, where the percentages refer to mass fractions of the bottled acids carrying a balance of H_2O. The first step is determining the mass fractions of the constituents:

$$w_{HNO_3} = 0.70 \times 0.76 = 0.532$$
$$w_{HF} = 0.49 \times 0.04 = 0.0196$$
$$w_{H_2O} = 1 - w_{HNO_3} - w_{HF} = 0.448$$

The approximate molar masses in kg/mol are

$$M_{HNO_3} = 0.063\ \mathrm{kg/mol}, \quad M_{HF} = 0.020\ \mathrm{kg/mol},\ M_{H_2O} = 0.018\ \mathrm{kg/mol}.$$

First derive the molality of the solvent, in mol/kg,

$$b_{H_2O} = \frac{1}{M_{H_2O}} = \frac{1}{0.018}\ \mathrm{mol/kg},$$

and use that to derive all the others by use of the equal ratios:

$$\frac{b_{HNO_3}}{b_{H_2O}} = \frac{w_{HNO_3} M_{H_2O}}{w_{H_2O} M_{HNO_3}} \quad \therefore b_{HNO_3} = 18.83\ \mathrm{mol/kg}.$$

Actually, b_{H2O} cancels out, because it is not needed. In this case, there is a more direct equation: we use it to derive the molality of HF:

$$b_{HF} = \frac{w_{HF}}{w_{H_2O} M_{HF}} = 2.19\ \mathrm{mol/kg}.$$

The mole fractions may be derived from this result:

$$x_{H_2O} = \frac{1}{1 + M_{H_2O}\left(b_{HNO_3} + b_{HF}\right)} = 0.726,$$
$$\frac{x_{HNO_3}}{x_{H_2O}} = \frac{b_{HNO_3}}{b_{H_2O}} \quad \therefore x_{HNO_3} = 0.246,$$
$$x_{HF} = 1 - x_{HNO_3} - x_{H_2O} = 0.029.$$

Osmolality

Osmolality is a variation of molality that takes into account only solutes that contribute to a solution's osmotic pressure. It is measured in osmoles of the solute per kilogram of water. This unit is frequently used in medical laboratory results in place of osmolarity, because it can be measured simply by depression of the freezing point of a solution, or cryoscopy (see also: osmostat and colligative properties).

Relation to Apparent (Molar) Properties

Molality appears in the expression of the apparent (molar) volume of a solute as a function of the molality b of that solute (and density of the solution and solvent):

$$^{\phi}\tilde{V}_1 = \frac{1}{b}\left(\frac{1}{\rho} - \frac{1}{\rho_0^0}\right) + \frac{M_1}{\rho}$$

For multicomponent systems the relation is slightly modified by the sum of molalities of solutes.

$$^{\phi}\tilde{V}_i = \frac{1}{\sum b_j}\left(\frac{1}{\rho} - \frac{1}{\rho_0^0}\right) + \frac{\sum b_j M_j}{\sum b_j \rho}$$

MOLAR CONCENTRATION

Molar concentration, also called molarity, amount concentration or substance concentration, is a measure of the concentration of a solute in a solution, or of any chemical species, in terms of amount of substance in a given volume. A commonly used unit for molar concentration used in chemistry is mol/L. A solution of concentration 1 mol/L is also denoted as 1 molar (1 M).

Definition

Molar concentration or molarity is most commonly expressed in units of moles of solute per litre of solution. For use in broader applications, it is defined as amount of solute per unit volume of solution, or per unit volume available to the species, represented by lowercase c:

$$c = \frac{n}{V} = \frac{N}{N_A\, V} = \frac{C}{N_A}.$$

Here, n is the amount of the solute in moles, N is the number of molecules present in the volume V (in litres), the ratio N/V is the number concentration C, and N_A is the Avogadro constant, approximately 6.022×10^{23} mol^{-1}. Or more simply: 1 molar = 1 M = 1 mole/litre.

In thermodynamics the use of molar concentration is often not convenient, because the volume of most solutions slightly depends on temperature due to thermal expansion. This problem is usually resolved by introducing temperature correction factors, or by using a temperature-independent measure of concentration such as molality.

The reciprocal quantity represents the dilution (volume) which can appear in Ostwald's law of dilution.

Units

In the International System of Units (SI) the base unit for molar concentration is mol/m^3. However, this is impractical for most laboratory purposes and most chemical literature traditionally uses mol/dm^3, or $moldm^{-3}$, which is the same as mol/L. These traditional units are often denoted by a capital letter M (pronounced *molar*), sometimes preceded by an SI prefix to denote sub-multiples, for example:

$mol/m^3 = 10^{-3}\ mol/dm^3 = 10^{-3}\ mol/L = 10^{-3}\ M = 1\ mmol/L = 1\ mM.$

The words "millimolar" and "micromolar" refer to mM and μM (10^{-3} mol/L and 10^{-6} mol/L), respectively.

Name	Abbreviation	Concentration	Concentration (SI unit)
millimolar	mM	10^{-3} mol/dm^3	10^{0} mol/m^3
micromolar	μM	10^{-6} mol/dm^3	10^{-3} mol/m^3
nanomolar	nM	10^{-9} mol/dm^3	10^{-6} mol/m^3
picomolar	pM	10^{-12} mol/dm^3	10^{-9} mol/m^3
femtomolar	fM	10^{-15} mol/dm^3	10^{-12} mol/m^3
attomolar	aM	10^{-18} mol/dm^3	10^{-15} mol/m^3
zeptomolar	zM	10^{-21} mol/dm^3	10^{-18} mol/m^3
yoctomolar	yM[3]	10^{-24} mol/dm^3 (1 particle per 1.6 L)	10^{-21} mol/m^3

Related Quantities

Number Concentration

The conversion to number concentration C_i is given by:

$$C_i = c_i \cdot N_A$$

where N_A is the Avogadro constant, approximately $6.022\times10^{23}\ mol^{-1}$

Mass Concentration

The conversion to mass concentration ρ_i is given by:

$$\rho_i = c_i \cdot M_i$$

where M_i is the molar mass of constituent i.

Mole Fraction

The conversion to mole fraction x_i is given by:

$$x_i = c_i \cdot \frac{M}{\rho} = c_i \cdot \frac{\sum_i x_i M_i}{\rho}$$

$$x_i = c_i \cdot \frac{\sum x_j M_j}{\rho - c_i M_i}$$

Where M is the average molar mass of the solution, ρ is the density of the solution and j is the index of other solutes.

A simpler relation can be obtained by considering the total molar concentration namely the sum of molar concentrations of all the components of the mixture.

$$x_i = \frac{c_i}{c} = \frac{c_i}{\sum c_i}$$

Mass Fraction

The conversion to mass fraction w_i is given by:

$$w_i = c_i \cdot \frac{M_i}{\rho}$$

Molality

The conversion to molality (for binary mixtures) is

$$b_2 = \frac{c_2}{\rho - c_2 \cdot M_2}$$

where the solute is assigned the subscript 2.

For solutions with more than one solute, the conversion is:

$$b_i = \frac{c_i}{\rho - \sum c_i \cdot M_i}$$

Properties

Sum of Molar Concentrations – Normalizing Relations

The sum of molar concentrations gives the total molar concentration, namely the density of the mixture divided by the molar mass of the mixture or by another name the reciprocal of the molar volume of the mixture. In an ionic solution, ionic strength is proportional to the sum of molar concentration of salts.

Sum of Products Molar Concentrations-Partial Molar Volumes

The sum of products between these quantities equals one.

$$\sum_i c_i \cdot \bar{V}_i = 1$$

Dependence on Volume

Molar concentration depends on the variation of the volume of the solution due mainly to thermal expansion. On small intervals of temperature the dependence is :

$$c_i = \frac{c_{i,T_0}}{(1 + \alpha \cdot \Delta T)}$$

where c_{i,T_0} is the molar concentration at a reference temperature, α is the thermal expansion coefficient of the mixture.

Spatial Variation and Diffusion

Molar and mass concentration have different values in space where diffusion happens.

Examples

Example 1: Consider 11.6 g of NaCl dissolved in 100 g of water. The final mass concentration ρ(NaCl) will be:

ρ(NaCl) = 11.6 g / (11.6 g + 100 g) = 0.104 g/g = 10.4 %

The density of such a solution is 1.07 g/mL, thus its volume will be:

V = (11.6 g + 100 g) / (1.07 g/mL) = 104.3 mL

The molar concentration of NaCl in the solution is therefore:

c(NaCl) = (11.6 g / 58 g/mol) / 104.3 mL = 0.00192 mol/mL = 1.92 mol/L

Here, 58 g/mol is the molar mass of NaCl.

Example 2: Another typical task in chemistry is the preparation of 100 mL (= 0.1 L) of a 2 mol/L solution of NaCl in water. The mass of salt needed is:

$m(\text{NaCl}) = 2 \text{ mol/L} \times 0.1 \text{ L} \times 58 \text{ g/mol} = 11.6 \text{ g}$

To create the solution, 11.6 g NaCl are placed in a volumetric flask, dissolved in some water, then followed by the addition of more water until the total volume reaches 100 mL.

Example 3: The density of water is approximately 1000 g/L and its molar mass is 18.02 g/mol (or 1/18.02=0.055 mol/g). Therefore, the molar concentration of water is:

$c(H_2O) = 1000 \text{ g/L} / (18.02 \text{ g/mol}) = 55.5 \text{ mol/L}$

Likewise, the concentration of solid hydrogen (molar mass = 2.02 g/mol) is:

$c(H_2) = 88 \text{ g/L} / (2.02 \text{ g/mol}) = 43.7 \text{ mol/L}$

The concentration of pure osmium tetroxide (molar mass = 254.23 g/mol) is:

$c(OsO_4) = 5.1 \text{ kg/L} / (254.23 \text{ g/mol}) = 20.1 \text{ mol/L}$.

Example 4: A typical protein in bacteria, such as *E. coli*, may have about 60 copies, and the volume of a bacterium is about 10^{-15} L. Thus, the number concentration C is:

$C = 60 / (10^{-15} \text{ L}) = 6 \times 10^{16} \text{ L}^{-1}$

The molar concentration is:

$$c = \frac{C}{N_A} = 6 \times 10^{16} \text{ L}^{-1} / (6 \times 10^{23} \text{ mol}^{-1}) = 10^{-7} \text{ mol/L} = 100 \text{ nmol/L}$$

Formal Concentration

If the concentration refers to original chemical formula in solution, the molar concentration is sometimes called *formal concentration*. For example, if a sodium carbonate solution has a formal concentration of $c(Na_2CO_3) = 1$ mol/L, the molar concentrations are $c(Na^+)$ = 2 mol/L and $c(CO_2{-}3)$ = 1 mol/L because the salt dissociates into these ions.

EQUIVALENT CONCENTRATION

In chemistry, the **equivalent concentration** or **normality** of a solution is defined as the molar concentration c_i divided by an equivalence factor f_{eq}:

$$\text{Normality} = \frac{c_i}{f_{eq}}$$

Unit Symbol N

The unit symbol "N" is used to denote "eq/L" (equivalent per liter) which is normality. Although losing favor, medical reporting of serum concentrations in "meq/L" (= 0.001 N) still occurs.

Usage

There are three common areas where normality is used as a measure of reactive species in solution:

- In acid-base chemistry, normality is used to express the concentration of hydronium ions (H_3O^+) or hydroxide ions (OH^-) in a solution. Here, $\frac{1}{f_{eq}}$ is an integer value. Each solute can produce one or more equivalents of reactive species when dissolved.
- In redox reactions, the equivalence factor describes the number of electrons that an oxidizing or reducing agent can accept or donate. Here, $\frac{1}{f_{eq}}$ can have a fractional (non-integer) value.
- In precipitation reactions, the equivalence factor measures the number of ions which will precipitate in a given reaction. Here, $\frac{1}{f_{eq}}$ is an integer value.

Normal concentration of an ionic solution is intrinsically connected to the conductivity (electrolytic) through the equivalent conductivity.

Examples

Normality can be used for acid-base titrations. For example, sulfuric acid (H_2SO_4) is a diprotic acid. Since only 0.5 mol of H_2SO_4 are needed to neutralize 1 mol of OH^-, the equivalence factor is:

$f_{eq}(H_2SO_4) = 0.5$

If the concentration of a sulphuric acid solution is $c(H_2SO_4) = 1$ mol/L, then its normality is 2 N. It can also be called a "2 normal" solution.

Similarly, for a solution with $c(H_3PO_4) = 1$ mol/L, the normality is 3 N because phosphoric acid contains 3 acidic H atoms.

Criticism

Normality is an ambiguous measure of the concentration of a solution. It needs a definition of the equivalence factor, which depends on the definition of equivalents. The same solution can possess *different* normalities for *different* reactions. The definition of the equivalence factor varies depending on the type of chemical reaction that is discussed: It may refer to equations, bases, redox species, precipitating ions, or isotopes. For example, a solution of $MgCl_2$ that is 2 N with respect to a Cl^- ion, is only 1 N with respect to an Mg^{2+} ion. Since f_{eq} may not be unequivocal, IUPAC and NIST discourage the use of normality.

REFERENCES

1. IUPAC, Compendium of Chemical Terminology, 2nd ed. (the "Gold Book") (1997). Online corrected version: (2006–) "amount fraction".

2. Zumdahl, Steven S. (2008). Chemistry (8th ed.). Cengage Learning. p. 201. ISBN 0-547-12532-1.
3. Rickard, James N.; Spencer, George M.; Bodner, Lyman H. (2010). Chemistry: Structure and Dynamics (5th ed.). Hoboken, N.J.: Wiley. p. 357. ISBN 978-0-470-58711-9.
4. Thompson, A.; Taylor, B. N. "The NIST Guide for the use of the International System of Units". National Institute of Standards and Technology. Retrieved 5 July 2014.
5. IUPAC, Compendium of Chemical Terminology, 2nd ed. (the "Gold Book") (1997). Online corrected version: (2006–) "volume fraction".
6. Volume-volume percentage, Chembuddy website
7. IUPAC, Compendium of Chemical Terminology, 2nd ed. (the "Gold Book") (1997). Online corrected version: (2006–) "mass fraction".
8. Formula from Mass Composition
9. IUPAC, Compendium of Chemical Terminology, 2nd ed. (the "Gold Book") (1997). Online corrected version: (2006–) "molality".
10. Journal of Solution Chemistry 5 (1976), 575
11. www.OED.com. Oxford University Press. 2011.
12. "NIST Guide to SI Units". Retrieved 2007-12-17.

CHAPTER 9

Intermolecular Force

Chapter Contents

- **Introduction**

Intermolecular forces are the attractive forces between molecules that hold them together. There are four types of intermolecular forces: hydrogen bonds and ionic, dipole and induced dipole forces.

INTRODUCTION

Intermolecular forces (IMFs) are forces of attraction or repulsion which act between neighboring particles (atoms, molecules, or ions). They are weak compared to the intramolecular forces, the forces which keep a molecule together. For example the covalent bond, involving the sharing of electron pairs between atoms is much stronger than the forces present between the neighboring molecules. They are an essential part of force fields frequently used in molecular mechanics.

The investigation of intermolecular forces starts from macroscopic observations which point out the existence and action of forces at a molecular level. These observations include non-ideal-gas thermodynamic behavior reflected by virial coefficients, vapor pressure, viscosity, superficial tension and absorption data.

The first reference to the nature of microscopic forces is found in Alexis Clairaut's work *Theorie de la Figure de la Terre*. Other scientists who have contributed to the investigation of microscopic forces include: Laplace, Gauss, Maxwell and Boltzmann.

Attractive intermolecular forces are considered by the following types:

- Ion-induced dipole forces
- Ion-dipole forces
- van der Waals forces (Keesom force, Debye force, and London dispersion force)

Information on intermolecular force is obtained by macroscopic measurements of properties like viscosity, PVT data. The link to microscopic aspects is given by virial coefficients and Lennard-Jones potentials.

Dipole-Dipole Interactions

Dipole-dipole interactions are electrostatic interactions between permanent dipoles in molecules. These interactions tend to align the molecules to increase attraction (reducing potential energy). An example of a dipole-dipole interaction can be seen in hydrogen chloride (HCl): the positive end of a polar molecule will attract the negative end of the other molecule and influence its position. Polar molecules have a net attraction between them. Examples of polar molecules include hydrogen chloride (HCl) and chloroform ($CHCl_3$).

$$\overset{\delta+}{\mathrm{H}} - \overset{\delta-}{\mathrm{Cl}} \cdots \overset{\delta+}{\mathrm{H}} - \overset{\delta-}{\mathrm{Cl}}$$

Often molecules contain dipolar groups, but have no overall dipole moment. This occurs if there is symmetry within the molecule that causes the dipoles to cancel each other out. This occurs in molecules such as tetrachloromethane and carbon dioxide. Note that the dipole-dipole interaction between two individual atoms is usually zero, since atoms rarely carry a permanent dipole.

Ion-Dipole and Ion-Induced Dipole Forces

Ion-dipole and ion-induced dipole forces are similar to dipole-dipole and induced-dipole interactions but involve ions, instead of only polar and non-polar molecules. Ion-dipole and ion-induced dipole forces are stronger than dipole-dipole interactions because the charge of any ion is much greater than the charge of a dipole moment. Ion-dipole bonding is stronger than hydrogen bonding.

An ion-dipole force consists of an ion and a polar molecule interacting. They align so that the positive and negative groups are next to one another, allowing for maximum attraction.

An ion-induced dipole force consists of an ion and a non-polar molecule interacting. Like a dipole-induced dipole force, the charge of the ion causes distortion of the electron cloud on the non-polar molecule.

Hydrogen Bonding

A hydrogen bond is the attraction between the lone pair of an electronegative atom and a hydrogen atom that is bonded to either nitrogen, oxygen, or fluorine. The hydrogen bond is often described as a strong electrostatic dipole-dipole interaction. However, it also has some features of covalent bonding: it is directional, stronger than a van der Waals interaction, produces interatomic distances shorter than the sum of van der Waals radius, and usually involves a limited number of interaction partners, which can be interpreted as a kind of valence.

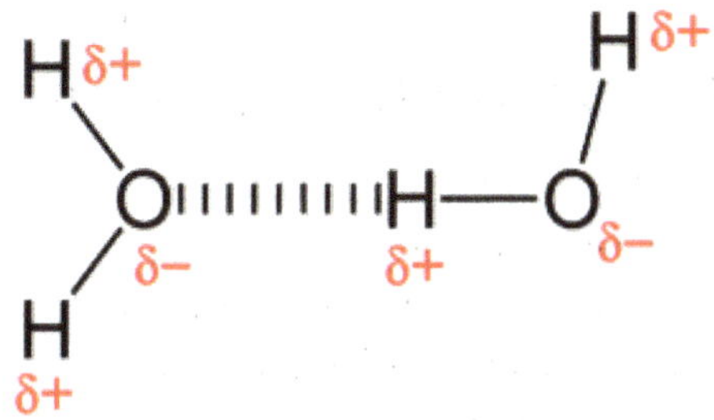

Intermolecular hydrogen bonding is responsible for the high boiling point of water (100 °C) compared to the other group 16 hydrides, which have no hydrogen bonds. Intramolecular hydrogen oxygen bonding is partly responsible for the secondary, tertiary, and quaternary structures of proteins and nucleic acids. It also plays an important role in the structure of polymers, both synthetic and natural.

Van der Waals forces

The van der Waals forces arise from interaction between uncharged atoms or molecules, leading not only to such phenomena as the cohesion of condensed phases and physical adsorption of gases, but also to a universal force of attraction between macroscopic bodies.

Keesom (Permanent-Permanent Dipoles) Interaction

The first contribution to van der Waals forces is due to electrostatic interactions between charges (in molecular ions), dipoles (for polar molecules), quadrupoles (all molecules with symmetry lower than cubic), and permanent multipoles. It is referred to as Keesom interactions(named after Willem Hendrik Keesom). These forces originate from the attraction between permanent dipoles (dipolar molecules) and are temperature dependent.

They consist of attractive interactions between dipoles that are ensemble averaged over different rotational orientations of the dipoles. It is assumed that the molecules are constantly rotating and never get locked into place. This is a good assumption, but at some point molecules do get locked into place. The energy of a Keesom interaction depends on the inverse sixth power of the distance, unlike the interaction energy of two spatially fixed dipoles, which depends on the inverse third power of the distance. The Keesom interaction can only occur among molecules that possess permanent dipole moments a.k.a. two polar molecules. Also Keesom interactions are very weak van der Waals interactions and do not occur in aqueous solutions that contain electrolytes. The angle averaged interaction is given by the following equation:

$$\frac{-m_1^2 m_2^2}{24\pi^2 \varepsilon_o^2 \varepsilon_r^2 k_b T r^6} = V$$

Where m = charge per length, ε_o = permitivity of free space, ε_r = dielectric constant of surrounding material, T = temperature, k_b = Boltzmann constant, and r = distance between molecules.

Debye (Permanent-Induced Dipoles) Force

The second contribution is the induction (also known as polarization) or Debye force, arising from interactions between rotating permanent dipoles and from the polarizability of atoms and molecules (induced dipoles). These induced dipoles occur when one molecule with a permanent dipole repels another molecule's electrons. A molecule with permanent dipole can induce a dipole in a similar neighboring molecule and cause mutual attraction. Debye forces cannot occur between atoms. The forces between induced and permanent dipoles are not as temperature dependent as Keesom interactions because the induced dipole is free to shift and rotate around the non-polar molecule. The Debye induction effects and Keesom orientation effects are referred to as polar interactions.

The induced dipole forces appear from the induction (also known as polarization), which is the attractive interaction between a permanent multipole on one molecule with an induced (by the former di/multi-pole) multipole on another. This interaction is called the Debye force, named after Peter J.W. Debye.

One example of an induction-interaction between permanent dipole and induced dipole is the interaction between HCl and Ar. In this system, Ar experiences a dipole as its electrons are attracted (to the H side of HCl) or repelled (from the Cl side) by HCl. The angle averaged interaction is given by the following equation.

$$\frac{-m_1^2 \alpha_2}{16\pi^2 \varepsilon_o^2 \varepsilon_r^2 r^6} = V$$

Where α= polarizability

This kind of interaction can be expected between any polar molecule and non-polar/symmetrical molecule. The induction-interaction force is far weaker than dipole-dipole interaction, but stronger than the London dispersion force.

London Dispersion Force (Induced Dipole-Induced Dipoles Interaction)

The third and dominant contribution is the dispersion or London force (fluctuating dipole-induced dipole), which arises due to the non-zero instantaneous dipole moments of all atoms and molecules. Such polarization can be induced either by a polar molecule or by the repulsion of negatively charged electron clouds in non-polar molecules. Thus, London interactions are caused by random fluctuations of electron density in an electron cloud. An atom with a large number of electrons will have a greater associated London force than an atom with fewer electrons. The dispersion (London) force is the most important component because all materials are polarizable, whereas Keesom and Debye forces require permanent dipoles. The London interaction is universal and is present in atom-atom interactions as well. For various reasons, London interactions (dispersion) have been considered relevant for interactions between macroscopic bodies in condensed systems. Hamaker developed the theory of van der Waals between macroscopic bodies in 1937 and showed that the additivity of these interactions renders them considerably more long-range.

Relative Strength of Forces

Bond type	Dissociation energy (kcal/mol)[9]
Ionic Lattice Energy	250–4000 [10]
Covalent Bond Energy	30–260
Hydrogen Bonds	1–12 (about 5 in water)
Dipole–Dipole	0.5–2
London Dispersion Forces	<1 to 15 (estimated from the enthalpies of vaporization of hydrocarbons)[11]

Note: this comparison is only approximate – the actual relative strengths will vary depending on the molecules involved. Ionic and covalent bonding will always be stronger than intermolecular forces in any given substance.

Effect on the Behavior of Gases

Intermolecular forces are repulsive at short distances and attractive at long distances (see the Lennard-Jones potential). In a gas, the repulsive force chiefly has the effect of keeping two molecules from occupying the same volume. This gives a real gas a tendency to occupy a larger volume than an ideal gas at the same temperature and pressure. The attractive force draws molecules closer together and gives a real gas a tendency to occupy a smaller volume than an ideal gas. Which interaction is more important depends on temperature and pressure (see compressibility factor).

In a gas, the distances between molecules are generally large, so intermolecular forces have only a small effect. The attractive force is not overcome by the repulsive force, but by the thermal energy of the molecules. Temperature is the measure of thermal energy, so increasing temperature reduces the influence of the attractive force. In contrast, the influence of the repulsive force is essentially unaffected by temperature.

When a gas is compressed to increase its density, the influence of the attractive force increases. If the gas is made sufficiently dense, the attractions can become large enough to overcome the tendency of thermal motion to cause the molecules to spread out. Then the gas can condense to form a solid or liquid (i.e., a condensed phase). Lower temperature favors the formation of a condensed phase. In a condensed phase, there is very nearly a balance between the attractive and repulsive forces.

Quantum Mechanical Theories

Intermolecular forces observed between atoms and molecules can be described phenomenologically as occurring between permanent and instantaneous dipoles, as outlined above. Alternatively, one may seek a fundamental, unifying theory that is able to explain the various types of interactions such as hydrogen bonding, van der Waals forces and dipole-dipole interactions. Typically, this is done by applying the ideas of quantum mechanics to molecules, and Rayleigh–Schrödinger perturbation theory has been especially effective in this regard. When applied to existing quantum chemistry methods, such a quantum mechanical explanation of intermolecular interactions, this provides an array of approximate methods that can be used to analyze intermolecular interactions.

REFERENCES

1. Margenau, H. and Kestner, N. (1969) Theory of intermolecular forces, International Series of Monographs in Natural Philosophy, Pergamon Press, ISBN 1483119289
2. Blaber, Michael (1996). Intermolecular Forces. mikeblaber.org
3. IUPAC, Compendium of Chemical Terminology, 2nd ed. (the "Gold Book") (1997). Online corrected version: (2006–) "hydrogen bond".
4. Leite, F. L.; Bueno, C. C.; Da Róz, A. L.; Ziemath, E. C.; Oliveira, O. N. (2012). "Theoretical Models for Surface Forces and Adhesion and Their Measurement Using Atomic Force Microscopy". International Journal of Molecular Sciences. 13 (12): 12773. doi:10.3390/ijms131012773.
5. Keesom, W.H. (1915). "The second virial coefficient for rigid spherical molecules whose mutual attraction is equivalent to that of a quadruplet placed at its center" (PDF). Proc. R. Acad. Sci. 18: 636–646.
6. Blustin, P. H. (1978). "A Floating Gaussian Orbital calculation on argon hydrochloride (Ar · HCl)". Theoretica Chimica Acta. 47 (3): 249. doi:10.1007/BF00577166.
7. ^ Jump up to:a b Roberts, J. K.; Orr, W. J. C. (1938). "Induced dipoles and the heat of adsorption of argon on ionic crystals". Transactions of the Faraday Society. 34: 1346. doi:10.1039/TF9383401346.
8. Sapse, A. M.; Rayez-Meaume, M. T.; Rayez, J. C.; Massa, L. J. (1979). "Ion-induced dipole H–n clusters". Nature. 278 (5702): 332. Bibcode:1979Natur.278..332S. doi:10.1038/278332a0.
9. Ege, Seyhan (2003) Organic Chemistry: Structure and Reactivity. Houghton Mifflin College. ISBN 0618318097. pp. 30–33, 67
10. "Lattice Energies". Retrieved 2014-01-21.
11. Majer, V. and Svoboda, V. (1985) Enthalpies of Vaporization of Organic Compounds, Blackwell Scientific Publications, Oxford. ISBN 0632015292

Index

A

Ancient philosophy posited 111
Atomic model 30
Atomic nucleus 9
Atomic number 94, 97, 98, 99, 100, 103, 104, 105, 107, 108, 109, 112, 114
Atomic theory 29
Atomistic position 41, 42

B

Baryonic matter 74

C

Characteristic properties 93
Chemical Abstracts Service (CAS) 91
Chemical bonds 127
Chemical composition 221, 222
Chemical compound 91, 92, 93, 94, 118
Chemical compounds 127
Chemical element 9, 92, 94, 99, 100, 112
Chemical equations 142
Chemical equilibrium 167
Chemical exchange 183
Chemical nature 11
Chemical properties 93, 97, 98, 99, 101, 103, 119
Chemical reaction 167, 192, 201
Chemical reactions 139, 140, 141, 144, 163
Chemical structure 92, 117, 119
Chemical substances 139, 141
chlorine (Cl) 15

D

deoxyribonucleic acid molecules (DNA) 70

E

Elementary reaction 143, 164
Exotic matter 79

F

Fermi–Dirac statistics 73, 74

G

Galaxy rotation curve 79

H

Hydrogen bond 242, 243, 246
hydrogen chloride (HCl) 242

I

Intermolecular forces 241, 245, 246
International System of Units (SI) 234
International Union of Pure and Applied Chemistry (IUPAC) 10, 104, 118, 120
Ion-dipole 242
Ionic bonds 129

K

kinetic theory 49, 50, 51, 52, 53, 54

L

linear combination of atomic orbitals molecular orbital method (LCAO) 130
London force 244

M

Magnetic materials 76
mass fraction 222, 223, 225, 226, 227, 229, 230, 235, 239
Metallic bonding 129, 130, 135
Modern physics 69
Molar concentration 233, 236

O

Ordinary matter 67, 68, 70, 71, 72, 73, 74, 77, 79
Organic chemistry 6, 23

P

Particle physics 73, 75, 77, 79, 82, 83
Percent concentration 226
Physical science 1
Planetary model 35

Q

Quantum chemistry 77
Quantum theory 36

R

Repulsive forces 246
Reverse reaction 151

S

Simultaneous measurement 183
Specific ion theory (SIT) 199
standard temperature and pressure (STP) 101, 102
Strong chemical bonds 131

T

Theoretical chemistry 23
Thermodynamic condition 171

V

Valence Shell Electron Pair Repulsion model (VSEPR) 15

W

Wilkinson Microwave Anisotropy Probe (WMAP) 74